T0140601

Springer Series in Fashion Business

Series editor

Tsan-Ming Choi, Institute of Textiles and Clothing, The Hong Kong Polytechnic University, Hung Hom, Kowloon, Hong Kong

This book series publishes monographs and edited volumes from leading scholars and established practitioners in the fashion business. Specific focus areas such as luxury fashion branding, fashion operations management, and fashion finance and economics, are covered in volumes published in the series. These perspectives of the fashion industry, one of the world's most important businesses, offer unique research contributions among business and economics researchers and practitioners. Given that the fashion industry has become global, highly dynamic, and green, the book series responds to calls for more in-depth research about it from commercial points of views, such as sourcing, manufacturing, and retailing. In addition, volumes published in Springer Series in Fashion Business explore deeply each part of the fashion industry's supply chain associated with the many other critical issues.

More information about this series at http://www.springer.com/series/15202

Sébastien Thomassey · Xianyi Zeng
Editors

Artificial Intelligence for Fashion Industry in the Big Data Era

 Springer

Editors
Sébastien Thomassey
ENSAIT-GEMTEX
Roubaix
France

Xianyi Zeng
ENSAIT-GEMTEX
Roubaix
France

ISSN 2366-8776   ISSN 2366-8784   (electronic)
Springer Series in Fashion Business
ISBN 978-981-13-4325-4   ISBN 978-981-13-0080-6   (eBook)
https://doi.org/10.1007/978-981-13-0080-6

Printed on acid-free paper

This Springer imprint is published by the registered company Springer Nature Singapore Pte Ltd. part of Springer Nature
The registered company address is: 152 Beach Road, #21-01/04 Gateway East, Singapore 189721, Singapore

Preface

In today's world, data have become one of the most valuable elements for society progress and industrial innovations. Supported by applications of the Internet, the big data environment has drastically changed our daily life and also the economic and business world.

The garment manufacturing, becoming fashion industry, is one of the oldest human activities and has come down through the centuries with continuously adapting to the technology and society advances. For the fashion industry, the big data era is very challenging but offers a huge scope of opportunities. This book deals with "fashion big data" which includes many types of data: point-of-sales (POS) data, geographic information systems (GIS) data, social media data, virtual 3D data, sensory data, textile physical data.

To manage and make a profitable use of these data, advanced techniques are required. Artificial Intelligence (AI) includes a set of techniques which are particularly suitable in such situation. Indeed, AI is able to deal with the "3V" of big data, namely Velocity, Variety, Volume with uncertainties, volatility, complexity in the fashion industry and related market. However, the implementation of these techniques is sometimes difficult and can scare some fashion companies.

Therefore, faced to the variety of methods and models, applications as well as data types, we propose this book, aiming to give an overview to practitioners and academics of the potential of AI methods in all the sectors of the fashion industry.

Artificial Intelligence for Fashion Industry in the Big Data Era offers through three parts: Part I—AI for Fashion Sales Forecasting, Part II—AI for Textile Apparel Manufacturing and Supply Chain, and Part III—AI for Garment Design and Comfort, 14 chapters written by 24 co-authors.

To be very specific, the topics covered in this volume are as follows:

- Introduction: Artificial Intelligence for Fashion Industry in the Big Data Era
- AI-Based Fashion Sales Forecasting Methods in Big Data Era
- Enhanced Predictive Models for Purchasing in the Fashion Field by Applying Regression Trees Equipped with Ordinal Logistic Regression
- A Data Mining-Based Framework for Multi-Item Markdown Optimization

- Social Media Analytics for Decision Support in Fashion Buying Processes
- Review of Artificial Intelligence Applications in Garment Manufacturing
- AI for Apparel Manufacturing in Big Data Era: A Focus on Cutting and Sewing
- A Discrete Event Simulation Model with Genetic Algorithm Optimisation for Customised Textile Production Scheduling
- An Intelligent Fashion Replenishment System Based on Data Analytics and Expert Judgment
- Blockchain-Based Secured Traceability System for Textile and Clothing Supply Chain
- Artificial Intelligence Applied to Multisensory Studies of Textile Products
- Evaluation of Fashion Design Using Artificial Intelligence Tools
- Garment Wearing Comfort Analysis Using Data Mining Technology
- Garment Fit Evaluation Using Machine Learning Technology

We hope that this book will provide valuable insights and will be greatly beneficial to the fashion business.

We gratefully acknowledge all the authors who have contributed to this book and all the anonymous reviewers for their essential works.

Finally, we would like to thank the Springer team for their kind support and patience during the building of this book project.

Roubaix, France   Sébastien Thomassey
February 2018   Xianyi Zeng

The original version of the book was revised: Incorrect co-author affiliation has been corrected. The erratum to this book is available at https://doi.org/10.1007/978-981-13-0080-6_15

Contents

Introduction: Artificial Intelligence for Fashion Industry in the Big Data Era 1
Sébastien Thomassey and Xianyi Zeng

Part I   AI for Fashion Sales Forecasting

AI-Based Fashion Sales Forecasting Methods in Big Data Era 9
Shuyun Ren, Chi-leung Patrick Hui and Tsun-ming Jason Choi

Enhanced Predictive Models for Purchasing in the Fashion Field by Applying Regression Trees Equipped with Ordinal Logistic Regression 27
Ali Fallah Tehrani and Diane Ahrens

A Data Mining-Based Framework for Multi-item Markdown Optimization 47
Ayhan Demiriz

Social Media Analytics for Decision Support in Fashion Buying Processes 71
Samaneh Beheshti-Kashi, Michael Lütjen and Klaus-Dieter Thoben

Part II   AI for Textile Apparel Manufacturing and Supply Chain

Review of Artificial Intelligence Applications in Garment Manufacturing 97
Radhia Abd Jelil

AI for Apparel Manufacturing in Big Data Era: A Focus on Cutting and Sewing 125
Yanni Xu, Sébastien Thomassey and Xianyi Zeng

**A Discrete Event Simulation Model with Genetic Algorithm
Optimisation for Customised Textile Production Scheduling**   153
Brahmadeep and Sébastien Thomassey

**An Intelligent Fashion Replenishment System Based on Data
Analytics and Expert Judgment**   173
Roberta Sirovich, Giuseppe Craparotta and Elena Marocco

**Blockchain-Based Secured Traceability System for Textile
and Clothing Supply Chain**   197
Tarun Kumar Agrawal, Ajay Sharma and Vijay Kumar

Part III AI for Garment Design and Comfort

**Artificial Intelligence Applied to Multisensory Studies of Textile
Products**   211
Zhebin Xue, Xianyi Zeng and Ludovic Koehl

Evaluation of Fashion Design Using Artificial Intelligence Tools   245
Yan Hong, Xianyi Zeng, Pascal Brunixaux and Yan Chen

**Garment Wearing Comfort Analysis Using Data
Mining Technology**   257
Kaixuan Liu

Garment Fit Evaluation Using Machine Learning Technology   273
Kaixuan Liu, Xianyi Zeng, Pascal Bruniaux, Xuyuan Tao, Edwin Kamalha
and Jianping Wang

**Erratum to: Artificial Intelligence for Fashion Industry
in the Big Data Era** E1
Sébastien Thomassey and Xianyi Zeng

Introduction: Artificial Intelligence for Fashion Industry in the Big Data Era

Sébastien Thomassey and Xianyi Zeng

Abstract With the emergence of the big data era, companies, and more especially fashion companies, are faced with a new relationship between consumers, suppliers, and competitors. Fashion companies have also to manage different data with many and complex correlations and dependencies between them and uncertainties related to human factors. It is crucial for companies to master these data flows to optimize their decision making. In such situations, artificial intelligent techniques are particularly efficient. The potential applications of artificial intelligence in fashion industry cover a wide scope from design support systems to fashion recommendation systems through sensory evaluation, intelligent tracking systems, textile quality control, fashion forecasting, decision making in supply chain management or social networks and fashion e-marketing. Thus, this book aims to illustrate the different possibilities and advantages of artificial intelligence for the fashion industry in the big data era. This introduction chapter provides a brief description of each chapter of this book.

Since decades, the fashion industry (clothing, footwear, accessories, jewels, etc.) keeps changing and requires enterprises' managers to continuously adapt their strategy to markets and technological innovations. Indeed, with the globalization, the fashion supply chain is becoming more and more complex to be controlled. With progressively increasing demands on personalization, the consumer is becoming more and more difficult to be satisfied. With the emergence of Internet, the competition and relationships between firms have been deeply changed, etc. In the same time, companies are mastering a huge quantity of data and wish to optimize their production and business activities by exploiting these data.

Indeed, the modern life is based on a connected mobile-oriented world, in which all data, generated by business transactions, physical sensors, social media networks, and other sources, continuously flow into all areas. This big data environment has dramatically changed all industrial sectors, including fashion industry. Especially, the fashion market will be deeply optimized by developing data-based tracking systems

---

S. Thomassey (✉) · X. Zeng
ENSAIT-GEMTEX, 2, allée Louise et Victor Champier, 59100 Roubaix, France
e-mail: sebastien.thomassey@ensait.fr

© Springer Nature Singapore Pte Ltd. 2018
S. Thomassey and X. Zeng (eds.), *Artificial Intelligence for Fashion Industry in the Big Data Era*, Springer Series in Fashion Business,
https://doi.org/10.1007/978-981-13-0080-6_1

1

from raw materials to finished products and shops, proposing data-based recommendations linking different production stages, design, and marketing services, generating new professional knowledge by learning from data, building data-based flexible manufacturing systems for small series productions, and exploiting new e-marketing methods.

However, managers and practitioners have to take into account a tremendous number of available data to optimize their activities. Thus, in order to deal with these massive data in a suitable way, intelligent techniques (fuzzy logic, neural networks, genetic algorithms, data fusion, clustering …) are efficient computational tools for data mining, knowledge representation, modeling, optimization, and decision making. These systems are especially suitable for the fashion industry since they can effectively deal with data with many and complex correlations and dependencies between them and uncertainties related to human factors, including sensory quality attributes, consumer behavior, designer's knowledge, and so on. Nevertheless, fashion companies do not widely use artificial intelligence (AI) techniques especially due to the following reasons: (1) the scope of these methods is still unknown; (2) the implementation and set up of AI algorithms on real data are too complicated; (3) the benefits cannot be clearly identified due to the lack of relevant business models; etc.

The scope of potential applications of AI in fashion industry is wide and has been intensively covered in the literature: Sensory Evaluation/Kansei Engineering (Zhu et al. 2010; Lu et al. 2013), Fashion Design support system (Wang et al. 2014; Mok et al. 2013), Fashion Recommendation Systems (Wang et al. 2015; Tu and Dong 2010), Modeling, Simulation and Optimization of Textile Processes (Yu et al. 2009; Veit 2012), Textile Quality Control (Bahlmann et al. 1999; Blaga and Dobrea 2009), Fashion Forecasting (Choi et al. 2014; Brahmadeep and Thomassey 2016), Decision Making in Textile/Apparel Supply Chain Management (Wong et al. 2013; Hui and Choi 2016), Intelligent Tracking Systems (Moon and Ngai 2008; Nayak et al. 2015), Social Networks and Fashion E-Marketing (Asur and Huberman 2010; Kim and Ko 2012),….

Motivated by the general industrial trend of performing data-driven operations with AI methods for the fashion industry, the aim of this book project is to give an update of AI techniques and their prospective implementations to respond to the current fashion industry challenges: from the design to the retailing through the manufacturing at all levels and the supplying. Thus, for each main sector of the fashion industry, this book provides a collection of peer-refereed papers which are a review of AI techniques and/or experimental works based on AI techniques illustrating the possibilities and advantages for the fashion industry. The book is organized in three parts: Part I—AI for fashion sales forecasting, Part II—AI for textile apparel manufacturing and supply chain, and Part III—AI for garment design and comfort. The following of this first chapter proposes a detailed introduction of these three parts.

Being composed of four chapters, the first part aims to offer an overview of potentials of AI and big data approaches for the fashion sales forecasting. Forecasting is crucial for companies in the fashion industry. The fashion market is also very volatile and makes the forecast very challenging. In this respect, the capabilities of AI are huge, especially in the current big data environment.

In Chap. 2, Shuyun Ren et al. give a comprehensive review of AI methods suitable for fashion forecasting in big data era and provide guidance on how AI methods can be implemented. The first part of the chapter is dedicated to describe the main AI techniques, their advantages and weakness. The new opportunities offered by big data environment are then pointed out. In this framework, a four-step process is proposed. For each step, a review of the most suitable techniques is achieved.

Ali Fallah Tehrani and Diane Ahrens propose a method to predict the magnitude of purchases from a catalogue of fashion products using a two-step procedure in Chap. 3. The aim of this work is to enhance the sales forecasting early in the season to enable the decision maker to buy the right quantity of products. The first step of the proposed method relies on an ordinal logistic regression which consists in the estimation, for each fashion item, of the probability of reaching a level of purchase between three: high, medium, or low. Regression trees are then used to refine the prediction and obtain a quantitative output. A case study is described to illustrate the benefit of the proposed method.

In Chap. 4, Ayhan Demiriz deals with forecasting weekly demand of retail items integrating pricing effects. His system is based on linear regression models which relies on items clustering into complementary or substitute product category. This multi-item dynamic pricing model outperforms the standard single-item pricing model. This chapter provides an perfect example of an intelligent support system that deals at the same time with forecast, optimization, and pricing issues in the fashion retailing.

In the last decades, social media has emerged as one of the most important opportunities for companies to enhance their consumer relationship. It also makes available a huge amount of data which can be successfully used to enhance the sales forecasting. Samaneh Beheshti-Kashi et al. propose in Chap. 5 a decision support system based on social analysis and evaluate the potential added value of social media analytics for fashion buying processes. A six-step text mining procedures are proposed to deal with both color detection and fashion topics tracking in social media. An experimental analysis is performed to illustrate the extraction of fashion topics and colors from fashion blogs.

The second part of this book includes five chapters, aiming at presenting one of the most important research themes in fashion industry: manufacturing and supply chain. Indeed, the complex and long manufacturing and supply processes require decision makers to master the most advanced analysis and optimization techniques. With the emergence of digital factories, tracking systems, POS data, GIS data, etc., more and more opportunities exist for enhancing the production and supply chains with various optimized processes. In this context, the AI techniques reach their full potential.

This part starts with two review chapters on the implementation of AI techniques in garment manufacturing. Radhia Abd Jelil proposes in Chap. 6 a systematic and comprehensive review of research articles related to the application of AI techniques in garment manufacturing. This review encompasses all the research issues encountered in garment manufacturing which are classified into three categories:

- production planning, control, and scheduling,
- garment quality control and inspection,
- garment quality evaluation.

In conclusion, the author tackles the main issues regarding the implementation of AI technologies in garment industry.

In Chap. 7, the literature review on the implementation of AI in apparel manufacturing by Yanni Xu et al. provides a detailed focus on the production planning, control, and scheduling and more especially on the cutting, sewing, finishing, and packaging processes. The chapter is structured to give a comprehensive analysis of the literature by manufacturing processes and also by AI methods. From a full awareness of the state of the art, perspectives of AI for apparel manufacturing in the big data era are proposed.

If the literature related to the AI in apparel manufacturing is very fruitful, the deployment of these techniques in textile/apparel industries is sparser. In Chap. 8, Brahmadeep et al. present an industrial case study where a hybrid system based on discrete event simulation and GA optimization enables to improve the production scheduling of customized textile products. In this study, the high variety of products, generated by the customization, and the production constraints make the formal mathematical modeling not possible. To increase the productiveness, the company also chooses to automate most of the manufacturing unit which leads to a decrease of the flexibility in the order planning. The proposed system permits to both model and simulate accurately the physical system and optimize the order scheduling to avoid delay and increase the "before time shipment."

In Chap. 9, Roberta Sirovich et al. describe and implement an intelligent replenishment system fashion stores. In most cases, fashion retailers are faced to the same issue: How to replenish my stores with a very few sales per SKU per week, a high volatility of demand, a limited store surfaces, huge variety of SKU, and a high number of stores? The proposed system deals with these constraints by integrating simultaneously the store manager's knowledge, an innovative inventory store budget which works as an internal marketplace and machine learning optimization. The system is evaluated in a real situation through a case study of an Italian retailer.

The globalization of the textile apparel supply chain has involved different unsustainable practices such as counterfeits, hazardous working conditions, sweatshop, ... To counter these issues, Tarun Kumar Agrawal et al. propose in Chap. 10 blockchain-based secured systems to enhance the tracking of fashion products. They demonstrate through a case example that how the implementation of the blockchain can create a better transparency, traceability, and accountability of the supply chain actors.

The third part of this book includes four chapters and is dedicated to the design and comfort of textile product. The garment design and comfort are the most relevant features from the consumer's point of view. Consequently, it is crucial for fashion companies to perfectly master these two factors. However, with the increase of e-shopping and virtual try-on and prototyping, the relationship with consumers has changed and requires new methods to evaluate, quantify, and develop the design and the comfort of garments.

The consumer's purchasing decision of a textile product is very complex to model since it not only calls on the sense of sight but also the touch. The Chap. 11 proposed by Zhebin Xue et al. deals with multi-sensory evaluation of textile products. In a first step, they develop a fuzzy and genetic algorithm to predict the tactile consumer perception of real textile product. Then, a systematic method based on rough inclusion degree and fuzzy inference systems is developed to integrate the visual interpretation of the product. An experimental study demonstrates that the proposed system is able to accurately predict the sensory human perception of textile products.

In Chap. 12, Yannick Hong et al. explain and develop an AI-based system for the evaluation of fashion design. This method relies on 3D design of garment which is the most suitable technique for personalized garment design. A personalized garment block is designed with controlled wearing ease. Then, a rule-based pattern modification system combined with a fuzzy-based evaluation system enables the fashion designer to quickly co-design the product with the consumer and thus ensures a high level of design satisfaction of consumers.

The physical try-on of product is very important in the garment choice and purchasing decision. This point remains the main concern of consumers when they buy a garment online. With the raise of digital technologies such 3D body scanners and 3D virtual software, new opportunities are offered to researchers to develop new systems for the virtual evaluation of wearing and comfort. In Chap. 13, Kaixuan Liu et al. present an approach to evaluate the wearing comfort dedicated to fashion designers. Their system relies on numerical 3D simulation of the clothing pressure distribution and is able to dynamically simulate 120 postures representing daily human activity. With data mining-based method, they provide a quantitative evaluation of the wearing comfort of a pant which enables designers to accurately know the parts to garment the more sensitive. Kaixuan Liu et al. investigate the garment fit evaluation issue in Chap. 14. They develop a machine learning system for automatic, rapid, and accurate garment fit evaluation to help consumers in shopping online. From the clothing digital pressures, the proposed system predicts a garment fit level which is related to the garment features and the consumer measurements. Two experiments both real and virtual are performed on 9 female subjects and 72 pants and demonstrate the ability of the system to work in e-shopping environment.

From this above introduction, it emerges that many significant topics are covered in this book. We believe that the content of this book will provide practitioners and academics theoretical/analytical methods and case studies to enable a better implementation of AI and a more valuable use of big data in fashion companies. We also hope that this book will be a significant inspiration for the future researches in fashion industry.

References

Asur S, Huberman BA (2010) Predicting the future with social. In: WI-IAT'10 proceedings of the 2010 IEEE/WIC/ACM international conference on web intelligence and intelligent agent technology, vol 1, pp 492–499

Bahlmann C, Heidemann G, Ritter H (1999) Artificial neural networks for automated quality control of textile seams. Pattern Recogn 32(6):1049–1060

Blaga M, Dobrea D (2009) Computer vision systems for textiles quality control. In: Proceedings of the 6th international conference on management of technological changes

Brahmadeep, Thomassey S (2016) Intelligent demand forecasting systems for fast fashion. In: Choi T-M (ed) Information systems for the fashion and apparel industry. Woodhead Publishing, pp 145–159

Choi TM, Hui CL, Yu Y (2014) Intelligent fashion forecasting systems: models and applications. Springer, Berlin, Heidelberg

Hui PCL, Choi T-M (2016) Using artificial neural networks to improve decision making in apparel supply chain systems. In: Choi T-M (ed) Information systems for the fashion and apparel industry. Woodhead Publishing, pp 97–107

Kim AJ, Ko E (2012) Do social media marketing activities enhance customer equity? An empirical study of luxury fashion brand. J Bus Res 65(10):1480–1486

Lu H, Chen Y, Du J (2013) An interactive system based on Kansei engineering to support clothing design process. Res J Appl Sci Eng Technol 6(24):4531–4535

Mok PY, Xu J, Wang XX, Fan JT, Kwok YL, Xin JH (2013) An IGA-based design support system for realistic and practical fashion designs. Comput Aided Des 45(11):1442–1458

Moon KL, Ngai EWT (2008) The adoption of RFID in fashion retailing: a business value-added framework. Indus Manag Data Syst 108(5):596–612

Nayak R, Singh A, Padhye R, Wang L (2015) RFID in textile and clothing manufacturing: technology and challenges. Fash Text 2:9

Tu Q, Dong L (2010) An intelligent personalized fashion recommendation system. In: International conference on communications, circuits and systems (ICCCAS). Chengdu, pp 479–485

Veit D (2012) Simulation in textile technology, theory and applications. Woodhead Publishing

Wang L, Zeng X, Koehl L, Chen Y (2014) A human perception-based fashion design support system for mass customization. In: Sun F, Li T, Li H (eds) Knowledge engineering and management. Advances in intelligent systems and computing, vol 214. Springer, Berlin, Heidelberg

Wang L, Xianyi Z, Koehl L, Chen Y (2015) Intelligent fashion recommender system: fuzzy logic in personalized garment design. IEEE Trans Hum Mach Syst 45:95–109

Wong C, Guo ZX, Leung SYS (2013) Optimizing decision making in the apparel supply chain using artificial intelligence (AI): from production to retail. Woodhead Publishing

Yu X, Guo S, Huang X (2009) Textile enterprise internal process simulation and optimization. In: International conference on e-business and information system security, EBISS'09, 23–24 May. Wuhan, China

Zhu Y, Zeng X, Koehl L, Lageat T, Charbonneau A, Chaigneau C (2010) A general methodology for analyzing fashion oriented textile products using sensory evaluation. Food Qual Prefer 21(8):1068–1076

Part I
AI for Fashion Sales Forecasting

AI-Based Fashion Sales Forecasting Methods in Big Data Era

Shuyun Ren, Chi-leung Patrick Hui and Tsun-ming Jason Choi

Abstract The demand of fashionable products is much difficult to be forecasted owing to the short-life cycle and high volatility driven by the ever-changing fashion trend. Many artificial intelligent (AI) methods and AI-based hybrid methods have been proven to be efficient for conducting fashion sales forecasting in previous studies. With the development and application of big data, information analytics would definitely lead to benefit for fashion sales forecasting, operation management, even the whole fashion supply chain coordination. However, few researches have studied the applicability of AI methods with big data. As we know, AI-based forecasting methods are time consuming and complex processing. In this chapter, we determine whether they are suitable and efficient for conducting fashion sales forecasting by high dimensional and large data. This paper aims to provide an up-to-date review on the commonly used and more efficient AI-based fashion sales forecasting methods and further examines the applicability of these methods in big data. How to make better use of these methods in big data era will also be conducted.

Keywords Artificial intelligent (AI) · Big data · Fashion industry · Sales forecasting

The original version of this chapter was revised: Incorrect co-author affiliation has been corrected. The erratum to this chapter is available at https://doi.org/10.1007/978-981-13-0080-6_15

S. Ren
Guangdong University of Technology, Guangzhou, China

C. Patrick Hui (✉) · T. Jason Choi
Institute of Textiles and Clothing, The Hong Kong Polytechnic University,
Hunghom, Kowloon, Hong Kong
e-mail: tchuip@polyu.edu.hk

© Springer Nature Singapore Pte Ltd. 2018
S. Thomassey and X. Zeng (eds.), *Artificial Intelligence for Fashion Industry in the Big Data Era*, Springer Series in Fashion Business,
https://doi.org/10.1007/978-981-13-0080-6_2

1　Introduction

Sales forecasting is a never stopping hot topic in academic research and industry practice. An accurate forecast of future demand will lead to proper retail inventory management, which is a fundamental part of operation management. Different from other traditional products, the fashion products featured as short-life cycle and the demand is highly volatile with the changing favorite of consumers and complex fashion trend (Choi et al. 2013; Choi and Sethi 2010) that makes the sales forecasting in fashion industry more challenging. In past decades, many kinds of forecasting methods, including statistic methods, AI methods, and combining methods, have been proposed and examined in both theoretical evidence and industry practice. Liu et al. (2013) and Nenni et al. (2013) conducted a detailed review of different kinds of general methods and explored the technical contents of these forecasting models. It indicates that statistical methods such as auto-regression, exponential smoothing, ARIMA, SARIMA were widely used techniques for fashion sales forecasting because of being fast, simple, well-explored, and easy to understand. With the advance of computing technologies, artificial intelligence (AI) models, AI-based methods both AI methods and some hybrid methods were suggested to be more efficient for conducting fashion sales forecasting than statistical methods. Artificial neural network (ANN) models (Sztandera 2004) and fuzzy logic models are the first kind of models being employed for fashion retail sales forecasting in the literatures. ANN-based methods perform well in terms of forecasting accuracy but require a quit long time to training model and complete the forecasting process. In fashion industry, special for fast fashion widely adopted by a number of international fashion companies such as Zara, Uniqlo, Mango and H&M in recent years, quick response to fashion trend and customer's preference is a key point for occupation of head start in the market. To achieve fast running speed, extreme learning machine (ELM)-based methods have been employed and the forecasting performance is proven to be more acceptable than ANN-based methods (Huang et al. 2006; Zhu et al. 2005). In past decades, besides forecasting accuracy, there are two important considering criteria to evaluate the forecasting methods in fashion industry: running speed and being able to conduct forecasting by limited amount of dataset. Nowadays, however, massive amount of data is available to conduct analysis and forecasting, with the development and widely application of big data and cloud computing. Big data contains large volumes of data represented in a variety of forms, and it keeps growing and making changes continuously. Plenty of the traditional sale forecasting methods proved to be efficient in solving small-scale datasets. But in terms of big data, these might not work, because it is difficult to structure large datasets and eliminate the noise at the same time. AI is one of the most efficient techniques to deal with challenges related to large datasets, so AI-based forecasting system is one of the partners for fashion sale forecasting under big data environment due to its fast computation speed and non-linear processing capability. With the consideration of the feature and challenge of fashion sale forecasting in big data era, this chapter conducts a comprehensive review in different AI-based fashion sales forecasting method to investigate if those methods are applicable to big data and how it works with other supplement tools. After that, a guidance will also be served on selecting the best predictions in different conditions to blend as much data, models, and methods as possible via post-processing to bring together information from multiple sources in order to improve the deterministic forecast as well as to quantify its uncertainty.

2 AI-Based Fashion Sales Forecasting Methods

Benefits from the advance of computer technology, AI forecasting methods which are more versatile than statistical models, have been widely employed to implement fashion sales forecasting (Frank et al. 2003; Au et al. 2008; El-Bakry and Mastorakis 2008; Yu et al. 2012). In this section, the forecasting advance and weakness of some popular AI methods in fashion sales forecasting are briefly reviewed and introduced.

2.1 ANN and ELM-Based Methods

Artificial neural network (ANN) methods which are able to provide satisfactory results in fashion sales forecasting are a popular set of forecasting models for predicting fashion product's sales. It is well known that even a simple ANN would take a substantial amount of time to complete a basic sales forecasting task [e.g., it may take several minutes, and evolutionary neural networks (ENNs) would even take a longer time (Au et al. 2008)]. It is well known that even a simple ANN would take a substantial amount of time to complete a forecasting task (e.g., it may take several minutes, and evolutionary neural networks (ENN) may take hours (Ren 2015). The long computational time becomes a major hurdle for the deployment of many ANN and ENN-based forecasting models in real-world fast fashion demand forecasting. Relatively recently, there is a proposal of a fast single-hidden layer feed-forward neural network (SLFN) called the extreme learning machine (ELM) (Sun et al. 2007, 2008; Hsu and Wang 2007; Huang et al. 2006; Rong et al. 2008; Xia et al. 2012). ELM is able to learn much faster than many conventional gradient-based learning methods reported in the classical neural networks literature. To the best of our knowledge, Sun et al. (2008) is the first piece of work which applies ELM in conducting fashion demand forecasting. In the following, we review this pioneering demand forecasting model. For more details (including detailed illustrations and figures), refer to Sun et al. (2008).

Despite the fact that ANN models perform well in terms of yielding high forecasting accuracy (as indicated by performance measures such as the mean-squared error), these forecasting models require a very long time to complete the forecasting task. In other words, they are very time consuming. The reason behind such a drawback comes from the fact that these models are all utilizing the gradient-based learning algorithms such as the backpropagation neural network (BPNN). To overcome this problem, the ELM-based models have emerged. However, ELM has its most critical drawback of being "unstable" as it can generate different outcomes in each different runs.

2.2   Fuzzy Logic-Based Methods

Fuzzy logic model proposed by Zadeh (1965) has been applied to conduct demand forecasting in many areas, such as Escoda et al. (1997), Chen and Wang (1999), Mastorocostas et al. (2001), Jandaghi et al. (2010), Chen and Chen (2011). In fashion industry, Vroman et al. (1998) are the pioneers to investigate fuzzy logic-based forecasting model. They suggest that fuzzy-based model is applicable for forecasting the sales of new fashion products of which the historical data are limited. Besides, benefits from the ability of identifying nonlinear relationships in input dataset, fuzzy logic-based forecasting model performs better in studying the effects of impact factors on future demand. Sztandera et al. (2004) developed a novel multivariate fuzzy model by using impact factors' data (including color, size), and it has been proved that the propose fuzzy-based model gets better performance compared with several statistical models such as Winters' three parameter exponential smoothing model (W3PES), the neural network, and the univariate forecasting models. Moreover, fuzzy logic-based model has good performance in fast fashion sales forecasting since it can reflect the inherent regular seasonality of demand, and it allows the input of expert knowledge Yesil et al. (2012). The merit of using fuzzy-based methods is that the technique can improve the efficiency of training and other processes with delicate design. However, the fuzzy techniques should be applied with other forecasting methods under a big data environment, such as ANN. Furthermore, its forecasting consumption time and efficiency are mainly decided by ANN or other AI methods. Thus, many researchers propose to combine fuzzy with other methods together to form a new "hybrid method" to achieve an efficient and effective forecasting task.

2.3   Support Vector Machines (SVMs)

As a member of machine learning approach, support vector machines (SVMs) have a very strong mathematical foundation. Thus, it has been previously applied to time series analysis Ruping and Morik (2003). Carbonneau et al. (2008) have examined the testing and forecasting performance of some machine learning methods (including NN, recurrent neural networks (RNNs) and SVM) on supply chain demand forecasting. The results show that SVM has the best performance on the training sets but performs poorly on the forecasting process because of the highest level forecasting error. Compared with traditional NN method, SVM can avoid the problems of excessive study, calamity data, and local minimal value, and so on by using use the theory of minimizing the structure risk (Wu 2009). Besides, the generalization property of

SVM does not depend on the complete training data but only a subset thereof. Due to these advance, SVM has been investigated to conduct forecasting problem in many studies, such as sales forecasting (Wu 2009), flood forecasting (Han et al. 2007), electricity load-price estimation (Swief et al. 2009), and so on. Moreover, SVM method has been used to conduct analysis based on large- and high-dimensional datasets (Wu 2009). Thus, in this chapter, the performance of SVM on big data forecasting process will also be carefully analyzed.

3 Application of Big Data in Fashion Industry

The advance of business intelligence (BI), defined as automatic data retrieving and processing systems that can help make intelligent decisions based on various data sources, has brought huge improvements to the efficiency of fashion supply chain management. The application of radio frequency identification (RFID) makes it possible to collected useful and large amount of data from retailers and suppliers in a supply chain. With the availability of real-time data from RFID, the accuracy of sales forecasting will be increased and the effectiveness of inventory management can thus increase. Besides, the Internet of things (IoTs), which promises machine–machine (M2M) communication, also provides a platform for collecting an enormous amount of data that can be transmitted to business intelligence and analytics tools for decisions making use. These big data collecting from either RFID or IoTs exactly records the sales information for each item or are linked to real-time events, thus, they can be analyzed to forecast the future demand.

RFID, as the most exciting and fastest-growing technology in terms of scope of application in the next generation of BI (Ngai et al. 1995), improves the limitation of real-time data collecting problem by traditional methods. It was used to lessen the bullwhip effect, improve inventory effectiveness even the supply chain efficiency in the literatures. Lee et al. (1997) use RFID to lessen the bullwhip effect by sharing and exchanging of supply chain information among manufacturers, suppliers, distributors, wholesalers, and retailers. In RFID-enabled supply chain system, decisions can be made at all supply chain levels (Saygin et al. 2007), so that the gap between the retail store and operations management is able to be well bridged as well. Following that, Mills-Harris et al. conducted a simulation study on the inventory management problem by using the mass data from RFID. Moreover, the improvement generated from RFID data-based analysis on inventory management even the whole supply chain performance has been investigated in many studies such as Visich et al. (2009), Sarac et al. (2010), Zelbst et al. (2010), Kok and Shang (2007). The benefit of RFID for retailers is that RFID records the data between products being received and products being sold in forms of a transponder ID, a reader ID, an event type, and a timestamp, called RFID events. These RFID events generated and stored in the company's information system are able to allow company to conduct several analyses of in-store processes, inventory status, and future demand forecasting.

Nowadays, billions of people around the world use the Internet for browsing the Web, shopping online, accessing retailers' Web site for information and services, communicating on social network, and so on. IoTs, as approach of converging data from different kinds of things, obtain mass data from the online platform on existing Internet infrastructure. It also can be looked at as a highly dynamic and radically distributed networked system, composed of a very large number of smart objects producing and consuming information (Miorandi et al. 2012). The data collected from IoTs sensors and devices consist not only of traditional discrete data, but also of streaming data. These mass data enable retailers accessing to the customs' behavior, the effects from big event on demand, and even other influence beyond our expectation.

However, in one hand, the raw data collected from RFID and IoTs, featured as vast, incomplete, noisy, obscure, and random, need to be processed by using data mining tools. Derakhshan et al. (2007) provide some usable methods. It extracts hidden information and knowledge which is useful and helpful for sales forecasting and analysis. On the other hand, the cloud service seems to be an effective technology that makes it possible to store and process this kind of big data originated from various locations and time points, and then data mining and cluster analyze (Choi et al. 2016).

4   AI-Based Fashion Sales Forecasting Methods in Big Data Era

As introduced above, big data is vast, incomplete, noisy, obscure, and even random compared with the small-scale datasets we used before. Although AI forecasting models are effective in solving the problem related to small-scale datasets, in terms of big data, they might not be useful to deal with large datasets. Big data is of large volume, complex, growing datasets which are generated from multiple autonomous sources (Wu et al. 2014). Because the data source is usually generated and collected without centralized control, several problems should be solved before predicting the trend and future. Therefore, AI-based technologies are applied not only in the forecasting system, but also solving the problems related to unstructured and semi-structured datasets that have noise.

Although research for fashion sales forecasting could not be found so far, it is well known that fashion sale related data usually contains a lot of features with small sample size, short lifecycle, and multidimensionality (Liu et al. 2013). These features lead to degradation in accuracy and efficiency of the forecasting system by curse of dimensionality. The cases are even worse in terms of big data structure, which leads to the degradation of classifier system performance in high-dimensional datasets because irrelevant features not only lead to insufficient classification accuracy, but also add extra difficulties in finding potentially useful knowledge. Therefore, a forecast of fashion sales system usually includes following steps to enable the effective forecasting of fashion sale (Fig. 1).

Step 1: In the first place, the noises of raw fashion sales related data should be filtered. Given the huge amounts of data are usually stored in a distributed fashion, it is highly imbalanced and contains a large amount of noises. Therefore, some method is needed to filter most noise data in the first place and leave the data of good quality to the next phase. If there are too much noise data, the forecasting results are certainly to be influenced and the training time of model will also increase to a great extent.

Step 2: Secondly, feature extraction method is applied to determine the relevant and irrelevant features of the dataset. It's a well known hard problem given the large-scale dataset and its complexity with so much noise. The filtering process assists in filtering the noises while some special optimizations should be used to extract useful data and its statistical features from the dataset before being used as the training data. There are two general approaches to feature extraction: filters and wrappers (Liu and Yu 2005). Filtered type methods extract features based on the intrinsic characteristics, which determine their relevance or discriminant power with regard to the target classes. Compared with wrapper-type methods, filter-type methods are relatively simpler. In wrapper-type methods, feature selection is "wrapped" around a learning method: The usefulness of a feature is directly judged by the estimated accuracy of the learning method. AI techniques usually benefit the wrapper-based methods through accelerating the learning speed or enhancing the estimation accuracy.

Step 3: After excluding irrelevant features and extracting useful features, AI-based machine learning is applied to model training and fitting. The training would be either in the form of unsupervised, supervised, or reinforcement. Note that the training data quality and the training method will significantly affect the prediction accuracy.

Step 4: Most of the forecasting systems are designed in a way of recursively improved. That is to say, the updated sales data of fashion products can be added into the training dataset to calibrate and improve the output accuracy continuously. Note that the output of the current iteration depends on the previous iteration of training and output optimization.

While AI-based method is the main scientific foundation of big data analytics, ANN-based, SVM-based, and ELM-based method are selected to introduce how AI-based model contributes in sale forecasting under big data circumstances.

Fig. 1 AI-based forecast process of fashion sales system

4.1 Data Filtering

When using big data, a problem we faced is that raw data are usually imbalanced with noises. In some cases, the number of training samples in some categories is much larger than that of other categories. In other cases, the data in different categories might have features of different scales. If the data were sent to training process with equal weights, small relative variations in the some categories make larger relative variations during the training, which would lead to an inefficient forecast. Last but not least, the disturbance from the noisy input data would affect the forecasting performance significantly.

4.1.1 ELM

Classical ELM sets all categories at the same scale, which treats all training data point equally. Weights can be given explicitly or generated automatically. Automatic weights generation assumes that the data have zero mean and unit variance. The weight information for each category is generated from the number of samples belonging to the same category. Therefore, data have features of different scales would have same influence. Some revised ELMs which incorporates different weights setting into ELM to improve the performance of handling the imbalanced data. Different from the classical one, Zong et al. (2013) added different penalty coefficients to the training errors corresponding to different inputs so as to re-weight the training data. The weighted ELM (W-ELM) leads to better results on imbalanced data by re-weighting training data to strengthen the impact of minority class while weakening the impact of majority class. Similarly, Mirza et al. (2013) introduced an improved tuning total error rate (TER) approach to assign weights for each category to processing the imbalanced data. Selecting input weights properly can also help in reducing the disturbance from the noisy input data. In, Man et al. (2011), a FIR-ELM is proposed to improve the performance of ELM on noise data, where the input weights are assigned based on the finite impulse response filter (FIR), such that the hidden layer performs as a preprocessor to improve the robustness of the model (Mirza et al. 2013). In, Man et al. (2012), a DFT-ELM is proposed based on discrete Fourier transform (DFT) technique. Similar to FIR-ELM, DFT-ELM also improves the robustness of ELM by designing the input weights to remove the effects of the disturbance from the noisy input data. These techniques show that ELM or advanced ELM can perform efficiency and superiority in handling imbalance input with noises in big data environment.

4.1.2 SVM

Classification technique is used to classify the unstructured data according to the format of the data that is to be processed, the analysis type to be applied, the data sources for the data that the target system is required to acquire, load, process, analyze, and store. If data are structured or linear, we can simply use maximum margin (Valizadegan and Jin 2006) or separation (Zimek et al. 2012) to classify the dataset. However, data are mostly unstructured/nonlinear in the era of big data. SVM-based technology is often used to filter the unstructured data into structured form by taking input from user.

Koturwar et al. (2015) developed a classification method combining SVM and K-NN. Items are classified by SVM first, the result obtained is further classified using K-NN to achieve more accurate results of classification. The filtered data from SVM and K-NN are taken to the learning module in the system to predict how to perform in the future. The kernel function used to perform the decision making is Sigmod (Lin and Lin 2003), which decides whether to send the data it gains from classifiers to weighted approximately ranked pairwise (WARP) or area under the curve (AUC) algorithms based on threshold value.

4.2 Feature Extraction

The data features are extracted from raw data in order to be converted to structured format. AI techniques not only help to reduce the dimensionality of large datasets but also to speed up the computation time of a learning algorithm and therefore simplify the classification task.

4.2.1 ELM

ELM usually benefits the feature extraction process when it uses wrapper procedure to select features out from the data. Usually, there exists a trade-off between the number of selected features and the generalization error. The problem is even much serious when the dataset is big data. More features means more info, while the curse of dimensionality and the finite number of samples available for learning may harm the idea view when too many features are considered. Another issue is that the best generalization error is usually not the only objective.

BenoîT et al. (2013) proposed a supervised feature extraction method based on ELM to make this trade-off. The main idea is that ELM with leave-one-out error (LOO) is used for fast evaluation of generalization error to guide the search for optimal feature subsets. It uses the gradient of the regularized training error of ELM to guide the search at each step, which only considers the direct neighbor pointed to by the gradient. This structure design provides not only the optimal number of features but also the evolution of the estimation of the generalisation error, and which features are selected for different number of selected features.

4.2.2 Fuzzy

The basic structure of the classic fuzzy inference system is a model that maps input characteristics to input membership functions, input membership function to rules, rules to a set of output characteristics, output characteristics to output membership functions, and the output membership function to a single-valued output, or a decision associated with the output (Azar 2010). Under big data environments, it is sometime very difficult to discern what the membership functions should look like simply by looking at data. Rather than choosing the parameters associated with a given membership function arbitrarily, these parameters could be chosen so as to tailor the membership functions to the input–output data in order to account for these types of variations in the data values.

Since its introduction, adaptive neuro-fuzzy inference system (ANFIS) networks (Atsalakis and Valavanis 2009) have been successfully applied to classification or feature extraction tasks. ANFIS is essentially an adaptive network functionally equivalent to FIS and structurally equivalent to an artificial neural network (ANN). It is able to construct an input–output mapping based on both human knowledge and stimulated input–output data pairs. Azar and Hassanien (2015) presented a fuzzy feature selection (FS) method based on the LH concept for classification of four medical datasets such as breast cancer Wisconsin (diagnostic), breast cancer Wisconsin prognostic, erythemato-squamous disease, and thyroid disease datasets. The study shows that the usage of neural-fuzzy classifier improves the performance of feature selection. It not only reduces the dimensions of the problem, but also improves classification performance by discarding redundant, noise-corrupted, or unimportant features and speed up the computation time of a learning algorithm and therefore simplifies the classification task.

4.2.3 SVM

Considering that SVM known as an effective classification model, it is useful to handle the feature extraction of big data. Yu et al. (2003) presented a method called clustering-based SVM (CB-SVM) which is suitable for feature extraction of big data. To perform the selective decluttering efficiently, the hierarchical structure constructs two clustering feature (CF) trees from positive and negative dataset independently. SVM boundary function can be trained from the root node of the two CF trees. Based on this structure, a single scan of the entire dataset would deliver statistical summaries that facilitate constructing a SVM boundary efficiently and effectively. The merit of using SVM-based methods to extract features is that the technique can select the data intelligently such that the degree of learning is maximized by the data.

4.2.4   ANN

Similar to ELM, ANN is a promising global searching approach for feature extraction. Au et al. (2008) employ ENN to search for the ideal network structure for a forecasting system, and then an ideal neural networks structure for fashion sales forecasting is developed. They reported that the performance of their proposed ENN model is better than the traditional SARIMA model for products with features of low-demand uncertainty and weak seasonal trends.

4.3   Data Training

Different from training a small data model in the working memory for quick access, the big data training algorithm relies on iterative processing of small chunks of data.

4.3.1   ELM

As introduced in Sect. 2.3, the random feature mapping in ELM ensures its universal approximation capability and makes it very efficient in training. Applications on intelligent high-dimensional and large data processing benefited a lot from ELM and its variants due to the significantly reduced computational complexity brought by the randomness in network parameters and the tuning-free learning strategy.

For an ELM with a large number of neurons trained on a huge dataset, almost all the running time is spent on training the correction matrices. However, through spiting the data into multiple parts, the computation can be conducted in parallel. The final correlation matrices can be combined according to its output weights. Chen et al. (2016) propose a MapReduce-based distributed ELM training framework named MR-ELM to enable large-scale and distributed ELM training on Hadoop cluster. Under big data environment, large volume sample blocks are located in different nodes of Hadoop cluster and can be accessed by Hadoop (Dittrich and Quiané-Ruiz 2012) file system. Through the framework of MapReduce, each Hadoop node trains an ELM sub-model—a group of hidden nodes of SLFN. Moving the training tasks to the sample blocks and collecting the trained hidden nodes cost much less I/O resources than collecting all the data blocks together for centralized training. Before forecasting by the final ELM model, the framework combines the sub-models as a complete ELM model. Because of MR-ELM's high scalability and efficiency, it can deal with big static sample data in a limited data with a Hadoop cluster. Comparing with conventional tuning-based parameter updating approaches, ELM has the advantages of extremely fast training speed and convincing generalization performance.

4.3.2   Fuzzy

Because of the ability of incorporating human knowledge and expertise for infer-
encing and decision making, fuzzy techniques are usually integrated to assist other
algorithm during the training process of forecasting. G. S. Atsalakis and K. P. Valavanis proposed a neuro-fuzzy system to forecasting stock market short-term trends
(Syed et al. 1999). Stock price dataset is a very big dataset; it is too expensive to
predict future absolute values of stocks on a daily basis, particularly since the stock
market is a memory less system. However, it is possible to appropriately train over any
(uptrend, downtrend, flat) horizon one has enough indicators to forecast trend with
significant accuracy via the help of adaptive neuro-fuzzy inference system (ANFIS).

In fashion retail sales forecasting, Sztandera et al. (2004) constructed a novel
multivariate fuzzy model which is based on several important product variables such
as color, time, and size. In their proposed model, grouped data and sales values
are calculated for each size-class combination. Compared with several statistical
models such as Winters' three parameter exponential smoothing model (W3PES),
the neural network model, and the univariate forecasting models, they find that their
proposed multivariable fuzzy logic model is an effective sales forecasting tool. In
fact, the good performance of the fuzzy logic-based models comes from their ability
to identify nonlinear relationships in the input data. In addition, the multivariate
fuzzy logic model performs better in comparison to the univariate counterparts. Later
on, Hui et al. (2005) explored the demand prediction problem in terms of fashion
color forecasting. They proposed a fuzzy logic system which integrates preliminary
knowledge of color prediction with the learning-based fuzzy color prediction system
to conduct forecasting. They reported several promising results of their proposed
method.

Actually, the fuzzy techniques are not only utilized in the training step but during the whole forecasting process. This methodology considers historical/past stock
prices as inputs of the fuzzy-based forecasting system that captures the underling
"laws of the stock market price motion" to predict next day's trend of a stock. Based
on the inverse learning technique also known as general learning (Atsalakis and
Valavanis 2009), the ANFIS is trained as a controller of the forecasting system to
determine the forecasting outputs. The proposed system has been proven that some
difficult forecasting mission is possible with the help of fuzzy techniques. The merit
of using fuzzy-based methods is that the technique can improve the efficiency of training and other process with delicate design. However, the fuzzy techniques should be
applied with other forecasting methods under a big data environment, such as ANN.
Furthermore, its forecasting consumption time and efficiency are mainly decided by
ANN or other AI methods.

4.3.3   SVM

Besides feature extraction, SVM is mainly adopted in the training step and forecasting
output step. Wang et al. (2013) proposed a hybrid method combing decision tree (DT)

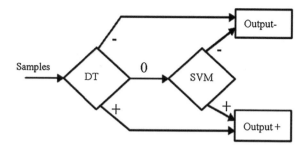

Fig. 2 Workflow of DT-SVM training

and support vector machine (SVM) method to train and forecast the price trends of the stock futures which are essential for investment decisions.

Raw data filtering and feature extraction have been applied to deal with huge amounts of futures data before training and forecasting. Firstly, the transaction data are obtained and filtered using a Hive (Sagiroglu and Sinanc 2013)-based distributed database. The data of each futures contract are split into different groups according to their time spans. Secondly, MapReduce-based method is applied to extract and compute the statistical features of the transaction data such as maximum, minimum, count, summation, mean value, sigma, median value, and median absolute deviation (MAD) of each group for different time spans simultaneously.

The extracted feature data are split by a certain time span and stored in different control nodes according to each futures contract. Considering the merit of tuning between precision and recall, DT is used to further filter and train. The precision rate of the result on each node is calculated simultaneously. Leaf nodes covering only positive samples are marked as "+" state; leaf nodes covering only negative nodes are marked as "−" state. Only the data on nodes with precision rate lower than the threshold are selected out to be stored in the output dataset and marked 0. These data would be further trained by SVM, and the trend of futures contracts in the next period of time would be calculated and output based on the results given by DT and SVM. In addition to the above two steps, the method sometimes might be improved recursively. In this case, SVM contributes little in this outputting step when the precision rate and recall rate are stabilized (Fig. 2).

4.3.4 ANN

The application of ANN model in training the fashion retail sales forecasting is similar to ELM. Frank et al. (2003) are one of the trailblazers to explore the use of ANN in fashion forecasting. It achieved better performance when compared it with two other statistical methods in terms of forecasting result. However, there are some problems which occur such as over learning due and over-fitting. Since ANN models can approximate essentially any function, they can also overfit all kinds of noise perfectly. Typically, fashion product's data are highly volatile with ever-changing taste of the consumers and the fashion product's life cycle is very short.

4.4 Forecast Output

It is quite common in big data classifications that some new training data arrived, some old training data expired, and some error training data corrected. When the dataset is big, the forecasting model usually suffers from excessive training time and complex parameters tuning problems, leading to inefficiency for real-time implementation and on-line model updating.

4.4.1 ELM

ELM has great potential for developing efficient and accurate forecasting model. It significantly improves the learning speed and enables effective on-line updating. Xu et al. (2013) developed an ELM-based predictor for real-time frequency stability assessment (FSA) of power systems. The inputs of the predictor are power system operational parameters, and the output is the frequency stability margin that measures the stability degree of the power system subject to a contingency. By off-line training with a frequency stability database, the predictor can be online applied for real-time FSA. The predictor was tested on New England 10-generator 39-bus test system, and the simulation results show that it can exactly accurately and efficiently predict the frequency stability. With the help of the universal mapping capability of ELM, Wan et al. (2014) developed an intelligent wind power prediction system based on ELM and the bootstrapping algorithm. Three different bootstrapping strategies, namely the pairs bootstrap, the standard residuals bootstrap, and the wild bootstrap, are employed to estimate model uncertainty intervals. This design enables ELM to generate the boundaries of the forecasting interval automatically, which reduces the complexity of optimizing decision variables.

4.4.2 SVM

On-line SVMs or incremental and decremental SVMs have been developed to handle dynamically incoming data efficiently (Cauwenberghs and Poggio 2000; Syed et al. 1999). In this scenario that an SVM model is incrementally constructed and maintained, the newer data have a higher impact on the SVM model than older data. It is because the data near the boundary have higher chances to be SVs in the next round of iteration. Thus, for the analysis of an archive data which should treat all the data equally, they would generate undesirable outputs.

With the help of other advanced algorithms and implantation techniques, SVM is developed as a much efficient tools in handling big data with higher speed. However, SVM-based methods need to process the entire dataset at every round to select/train/output the data, which generates too much I/O cost for very large datasets.

5 Conclusion

This chapter reviews the advance and weakness of some popular AI methods in fashion sales forecasting. Under big data environment, the demand of fashionable products is much difficult to be forecasted owing to the short-life cycle and high volatility driven by the ever-changing fashion trend. The commonly used ANN- and ELM-based methods, fuzzy-based methods, and SVM-based techniques have methods have been proven to be efficient for conducting fashion sales forecasting in previous studies.

Considering the features of big data, this chapter provides guidance on how commonly used AI methods benefit the four consecutive steps in fashion forecasting. In the data filtering step, ELM or advanced ELM can perform efficiency and superiority in handling imbalance input with noises in big data environment. SVM usually filters the unstructured data into structured form by taking input from user. During feature extraction, ELM usually benefits the extraction when it uses wrapper procedure to select features out from the data. With the assistance given by fuzzy techniques, it not only reduces the dimensions of the problem, but also improves classification performance by discarding redundant, noise-corrupted, or unimportant features and speed up the computation time of a learning algorithm and therefore simplifies the classification task. The merit of using SVM-based methods to extract features is that the technique can select the data intelligently such that the degree of learning is maximized by the data. Date training is a key process of fashion forecasting system; ELM has the advantages of extremely fast training speed and convincing generalization performance. Fuzzy techniques are usually integrated to assist other algorithms during the training process of forecasting. With the help of fuzzy techniques, some difficult forecasting task is achievable. Combing with other techniques, SVM also shows effective performance in big data training and forecasting. When the dataset is big, the forecasting model usually suffer from excessive training time and complex parameters tuning problems, leading to inefficiency for real-time implementation and on-line model updating. ELM is able to generate the boundaries of the forecasting interval automatically, which reduces the complexity of optimizing decision variables. Similar to the training steps, SVM can also assist other advanced techniques to handle big data with higher speed.

References

Atsalakis GS, Valavanis KP (2009) Forecasting stock market short-term trends using a neuro-fuzzy based methodology. Expert Syst Appl 36(7):10696–10707

Au K-F, Choi T-M, Yu Y (2008) Fashion retail forecasting by evolutionary neural networks. Int J Prod Econ 114(2):615–630

Azar AT (2010) Adaptive neuro-fuzzy systems. In: Fuzzy systems. In-Tech, Austria, 85–110

Azar AT, Hassanien AE (2015) Dimensionality reduction of medical big data using neural-fuzzy classifier. Soft Comput 19(4):1115–1127

BenoîT FN, Van Heeswijk M, Miche Y et al (2013) Feature selection for nonlinear models with extreme learning machines. Neurocomputing 102:111–124

Carbonneau R, Laframboise K, Vahidov R (2008) Application of machine learning techniques for supply chain demand forecasting. Eur J Oper Res 184(3):1140–1154

Cauwenberghs G, Poggio T (2000) Incremental and decremental support vector machine learning. In: Proceedings advances in neural information processing systems. Vancouver, Canada

Chen S-M, Chen C-D (2011) TAIEX forecasting based on fuzzy time series and fuzzy variation groups. IEEE Trans Fuzzy Syst 19(1):1–12

Chen T, Wang MJJ (1999) Forecasting methods using fuzzy concepts. Fuzzy Sets Syst 105(3):339–352

Chen J, Chen H, Wan X et al (2016) MR-ELM: a MapReduce-based framework for large-scale ELM training in big data era. Neural Comput Appl 27(1):101–110

Choi T-M, Sethi S (2010) Innovative quick responseprograms: a review. Int J Prod Econ 127(1):1–12

Choi TM, Hui CL, Yu Y (eds) (2013) Intelligent fashion forecasting systems: models and applications. Springer, New York, NY, USA

Choi TM, Chan HK, Yue X (2016) Recent development in big data analytics for business operations and risk management

Derakhshan R, Orlowska ME Li X (2007) RFID data management: challenges and opportunities. In: IEEE International conference on RFID March 2007, vol 10

Dittrich J, Quiané-Ruiz JA (2012) Efficient big data processing in Hadoop MapReduce. Proc VLDB Endow 5(12):2014–2015

El-Bakry HM, Mastorakis N (2008) A new fast forecasting technique using high speed neural networks. WSEAS Trans Signal Process 4(10):573–595

Escoda I, Ortega A, Sanz A, Herms A (1997) Demand forecast by neuro-fuzzy techniques. In: Proceedings of the 6th IEEE international conference on fussy systems (FUZZ-IEEE'97), July 1997. pp 1381–1386

Frank C, Garg A, Sztandera L, Raheja A (2003) Forecasting women's apparel sales using mathematical modeling. Int J Cloth Sci Technol 15(2):107–125

Han D, Chan L, Zhu N (2007) Flood forecasting using support vector machines. J Hydroinform 9(4):267–276

Hsu LC, Wang CH (2007) Forecasting the output of integrated circuit industry using a grey model improved by the Bayesian analysis. Technol Forecast Soc Chang 74(6):843–853

Huang G-B, Zhu Q-Y, Siew C-K (2006) Extreme learning machine: theory and applications. Neurocomputing 70(1–3):489–501

Hui C-L, Lau T-W, Ng S-F, Chan C-C (2005) Learningbased fuzzy colour prediction system for more effective apparel design. Int J Cloth Sci Technol 17(5):335–348

Jandaghi G, Tehrani R, Hosseinpour D, Gholipour R, Shadkam SAS (2010) Application of fuzzy-neural networks in multi-ahead forecast of stock price. Afr J Bus Manag 4(6):903–914

Kok AG, Shang KH (2007) Inspection and replenishment policies for systems with inventory record inaccuracy. Manuf Serv Oper Manag 9(2):185–205

Koturwar P, Girase S, Mukhopadhyay D (2015) A survey of classification techniques in the area of big data. arXiv:1503.07477

Lee HL, Padmanabhan V, Whang S (1997) The bullwhip effect in supply chain. Sloan Manag Rev 38(3):93–102

Lin HT, Lin CJ (2003) A study on sigmoid kernels for SVM and the training of non-PSD kernels by SMO-type methods. Neural Comput 1–32

Liu H, Yu L (2005) Toward integrating feature selection algorithms for classification and clustering. IEEE Trans Knowl Data Eng 17(4):491–502

Liu N, Ren S, Choi T-M, Hui C.-L, Ng S-F (2013) Sales forecasting for fashion retailing service industry: a review. Math Probl Eng. http://dx.doi.org/10.1155/2013/738675

Man ZH, Lee K, Wang DH, Cao ZW, Miao CY (2011) A new robust training algorithm for a class of single-hidden layer feedforward neural networks. Neurocomputing 74(16):2491–2501

Man ZH, Lee K, Wang DH, Cao ZW, Khoo SY (2012) Robust single-hidden layer feedforward network-based pattern classifier. IEEE Trans Neural Netw Learn Syst 23(12):1974–1986

Mastorocostas PA, Theocharis JB, Petridis VS (2001) A constrained orthogonal least-squares method for generating TSK fuzzy models: application to short-term load forecasting. Fuzzy Sets Syst 118(2):215–233

Miorandi D, Sicari S, De Pellegrini F, Chlamtac I (2012) Internet of things: vision, applications and research challenges. Ad Hoc Netw 10(7):1497–1516

Mirza B, Lin Z, Toh KA (2013) Weighted online sequential extreme learning machine for class imbalance learning. Neural Process Lett 38(3):465–486

Nenni ME, Giustiniano L, Pirolo L (2013) Demand forecasting in the fashion industry: a review. Int J Eng Bus Manag

Ngai EWT, Moon KKL, Riggins FJ, Yi CY (2008) RFID research: an academic literature review (1995–2005) and future research directions. Int J Prod Econ 112(2) 510–520

Ren S, Choi T-M, Liu N (2015) Fashion sales forecasting with a panel data-based particle-filter model. IEEE Trans Syst Man Cybern Syst 45(3), 411–421

Rong HJ, Ong YS, Tan AH, Zhu Z (2008) A fast pruned-extreme learning machine for classification problem. Neurocomputing 72(1–3):359–366

Ruping S, Morik K (2003) Support vector machines and learning about time. Paper presented at the Proceedings of ICASSP 2003

Sagiroglu S, Sinanc D (2013) Big data: a review. In: 2013 international conference on collaboration technologies and systems (CTS). IEEE, pp 42–47

Sarac A, Absi N, Dauzre-Prs S (2010) A literature review on the impact of RFID technologies on supply chain management. Int J Prod Econ 128(1):77–95

Saygin C, Sarangapani J, Grasman SE (2007) A systems approach to viable RFID implementation in the supply chain. In: Springer series in advanced manufacturing: trends in supply chain design and management technologies and methodologies vol 327

Sun ZL, Au KF, Choi TM (2007) A neuro-fuzzy inference system through integration of fuzzy logic and extreme learning machines. IEEE Trans Syst Man Cybern Part B Cybern 37(5):1321–1331

Sun ZL, Choi TM, Au KF, Yu Y (2008) Sales forecasting using extreme learning machine with applications in fashion retailing. Decis Support Syst 46(1):411–419

Swief RA, Hegazy YG, Abdel-Salam, TS Bader MA (2009) Support vector machines (SVM) based short term electricity load-price forecasting. In: 2009 IEEE bucharest powertech. pp 1–5

Syed N, Liu H, Sung K (1999) Incremental learning with support vector machines. In: Proceedings the workshop on support vector machines at the international joint conference on artificial intelligence. Stockholm, Sweden

Sztandera LM, Frank C, Vemulapali B (2004) Predicting women's apparel sales by sot computing. In: Proceedings of the 7th international conference on artificial intelligence and sot computing (ICAISC'04), June 2004. Zakopane, Poland, pp 1193–1198

Valizadegan H, Jin R (2006) Generalized maximum margin clustering and unsupervised kernel learning. In: Advances in neural information processing systems. pp 1417–1424

Visich JK, Li S, Khumawala BM, Reyes PM (2009) Empirical evidence of RFID impacts on supply chain performance. Int J Oper Prod Manag 29(12):1290–1315

Vroman P, Happiette M, Rabenasolo B (1998) Fuzzy adaptation of the Holt-Winter model for textile sales-forecasting. J Text Inst 89(1):78–89

Wan C, Xu Z, Pinson P et al (2014) Probabilistic forecasting of wind power generation using extreme learning machine. IEEE Trans Power Syst 29(3):1033–1044

Wang D, Liu X, Wang M (2013) A DT-SVM strategy for stock futures prediction with big data. In: 2013 IEEE 16th International Conference on computational science and engineering (CSE). IEEE, pp 1005–1012

Wu Q (2009) The forecasting model based on wavelet ν-support vector machine. Expert Syst Appl 36(4):7604–7610

Wu X, Zhu X, Wu GQ et al (2014) Data mining with big data. IEEE Trans Knowl Data Eng 26(1):97–107

Xia M, Zhang Y, Weng L, Ye X (2012) Fashion retailing forecasting based on extreme learning machine with adaptive metrics of inputs. Knowl-Based Syst 36:253–259

Xu Y, Dai Y, Dong ZY et al (2013) Extreme learning machine-based predictor for real-time frequency stability assessment of electric power systems. Neural Comput Appl 22(3–4):501–508

Yesil E, Kaya M, Siradag S (2012) Fuzzy forecast combiner design for fast fashion demand forecasting. In: Proceedings of the IEEE international symposium in innovations in intelligent systems and applications (INISTA'12). pp 1–5

Yu H, Yang J, Han J (2003) Classifying large data sets using SVMs with hierarchical clusters. In: Proceedings of the ninth ACM SIGKDD international conference on knowledge discovery and data mining. ACM, pp 306–315

Yu J, de Jong R, Lee L-F (2012) Estimation for spatial dynamic panel data with fixed effects: the case of spatial cointegration. J Econom 167(1):16–37

Zadeh LA (1965) Fuzzy sets. Inf Comput 8:338–353

Zelbst PJ, Green KW Jr, Sower VE, Baker G (2010) RFID utilization and information sharing: the impact on supply chain performance. J Bus Indus Market 25(8):582–589

Zhu Q-Y, Qin AK, Suganthan PN, Huang G-B (2005) Evolutionary extreme learning machine. Pattern Recogn 38(10):1759–1763

Zimek A, Schubert E, Kriegel HP (2012) A survey on unsupervised outlier detection in high-dimensional numerical data. Stat Anal Data Min 5(5):363–387

Zong W, Huang GB, Chen Y (2013) Weighted extreme learning machine for imbalance learning. Neurocomputing 101:229–242

Enhanced Predictive Models for Purchasing in the Fashion Field by Applying Regression Trees Equipped with Ordinal Logistic Regression

Ali Fallah Tehrani and Diane Ahrens

Abstract Identifying the products which are highly sold in the fashion apparel industry is one of the challenging tasks, which leads to reduce the write-off and increase the revenue. Assuming three classes as substantial, middle, and inconsiderable, the forecasting problem comes down to a classification problem, where the task is to predict the class of a product. In this research, we present a probabilistic approach to identify the class of fashion products in terms of sale. In previous work, we showed that a combination of kernel machines with a probabilistic approach may empower the performance of kernel machines. However, a well-known drawback of kernel machines is its non-interpretability. The interpretability is one of the most important features from an user point of view; essentially in the fashion field, decision makers require to understand and interpret the model for a more convenient adaptation. Since regression trees can be formulated through rules, this makes possible to comprehend the model. Nevertheless, a drawback of decision trees is the sensibility to input space, which may cause very enormous deviations in terms of prediction. To reduce this effect on forecast, we propose a new model equipped with ordinal logistic regression. Finally to verify the proposed approach, we conducted several experiments on a real data extracted from an apparel retailer in Germany.

Keywords Fashion products · Comprehensibility · Kernel machines
Ordinal logistic regression · Regression trees · Sales forecasting

A. Fallah Tehrani (✉) · D. Ahrens
Technology Campus Grafenau - Deggendorf Institute of Technology,
Hauptstraße 3, 94481 Grafenau, Germany
e-mail: ali.fallahtehrani@th-deg.de

D. Ahrens
e-mail: diane.ahrens@th-deg.de

© Springer Nature Singapore Pte Ltd. 2018
S. Thomassey and X. Zeng (eds.), *Artificial Intelligence for Fashion
Industry in the Big Data Era*, Springer Series in Fashion Business,
https://doi.org/10.1007/978-981-13-0080-6_3

1 Introduction

Recently applying data analytics techniques, e.g., machine learning methods on the use of business forecasting for tremendous and complex data have received considerable attention. Specifically for the goal sales forecasting, these techniques attempt to recognize established patterns in the existing sales history. To extract sound patterns from data, the forecasting models should take into account all deciding factors. Unlike expert knowledge-based system, machine learning approaches provide more reliable and objective results, since they handle the data in an objective manner. Needless to say that the experts in apparel industry use some established rules,[1] which often may not consider any change over time; however, from a customer point of view, the so-called customer's interest is being changed notably over time. Therefore, there exists a danger of a wrong conclusion particularly due to the fact that the experts consider typically the same patterns. In contrast to this, however, machine learning techniques aim to bias observations in a reliable and convenient manner.

Typically in the fashion apparel industries, ordering is performed once in each season; reordering usually is very costly and takes time, too. Seen from this view, an accurate ordering causes more benefits and also reduces write off and overproduction. To this end, identifying products which may sell well can reduce drastically the overproduction. Naturally, the whole procedure is affected mostly by highly sold products, which henceforth we refer them to *trendy products*. A trend is assigned to a fashion product when it exceeds a threshold in terms of sales. We discuss more in the detail the way to categorize the products on the basis of number of sales. Since the trendy products effect on the majority of sales, detecting them may lead to enhancing the prediction accuracy.

Ordinarily purchasing process is conducted before starting new season through a reasonable approximation of demand. To this end, the retailer sends catalogues to its customers, and based on their feedbacks, order sought articles from other upstream retailers, most of which are located in Asia. Therefore, an accurate sales-ordering is desirable and leads to decrease in unnecessary expenditures. Note that, the most advertised fashion products are distinctive from year to year. Seen from this view, the estimator is required to recognize the proper patterns from the sales history. Unlike common fashion retailers which make forecast on the basis of conventional factors such as forecasting trend products and trend colors, catwalk analysis report, materials and texture analysis, seasonality, climate factors, and promotion, this retailer focuses mainly on features derived from the catalogue. Specifically for our forecasting machine, we use the following factors: color, size of photograph, price, and the number of colors offered for a given article; colors are clustered into 4–6 different clusters, size of photograph is one of most important factors indicating sales impact, price is assumed without any promotion, and finally, the number of colors shows diversity of articles. While typically conventional fashion retailers require to make prediction for next week or next month, our goal is to approximate demands for a

---

[1]For instance, there have existed several established patterns for color combinations, which are no longer valid.

whole season. Note that it is not feasible to modify order quantity among the season; the ordering is performed once in a season. It is worthwhile noting that while too much ordering may lead to price reductions, tied-up capital, and resource destruction, a very low purchasing causes a large revenue loss and an increase in terms of customer exasperation.

This study aims at enhancing the quality of forecast in two ways: Firstly, it deals with forecasting underlying probabilistic approaches to predict (classify) the class of new fashion products and secondly, on the basis of the classes attempts to predict the number of sales for future. As is clear, the first challenge belongs to classification methodology, whereas the second challenge is counted as a regression problem. To this end, we propose a novel method for regression on the basis of regression tree equipped with the ordinal class classification setting. Moreover, we show that by employing our proposed predictive model there is a chance to overcome the obstacles (outliers and noises), which indeed outperforms common regression methods.

The rest of the paper is organized as follows: After a short survey on existing methodologies in the next section, in Sect. 3 ordinal logistic regression is presented. Section 4 introduces decision trees and Iterative Dichotomiser 3 algorithm. Section 5 is dedicated to the algorithm and our proposal. Finally, the results are demonstrated in Sect. 6.

2 Related Work

In this section, we explain approaches for the sales-prediction in fashion field: Thomassey and Happiette (2007) employ neural network method, specifically neural clustering and classification (NCC) to tackle the problem. Kit Wong and Guo (2010) apply a hybrid intelligent model underlying neural networks to make a prediction in terms of medium-term fashion. In this regard, Alon et al. (2001) serve a comparison between neural network methods and conventional methods for the goal prediction retail sales. To analyze sales dynamically, Frank et al. (2003) deal with the forecasting in a different manner. In fact, they proposed online-learning, which basically supposes daily lag and weekly lag. Then, they define on each lag autocorrelation function (ACF). Since feedbacks can be monitored dynamically, it is possible to improve the accuracy of prediction by neural network method. Apart from conventional neural network, Kuo et al. (2002) equip the common neural networks by fuzzy weights, called fuzzy neural networks. Huang and Qiurui (2017) combine fuzzy inference system with artificial neural network. This model contains two stages: In first stage, an estimator makes a prediction on the base of original data concerning stock-out, and in second stage, the adaptive neuro-fuzzy inference system refines final demands. Rodrigues and Figueiredo (2013) also apply artificial neural network on the use of demand forecast. Beyond neural network methods, Thomassey et al. (2002a, b) use classification methods to cope with mean-term forecasting. Happiette et al. (1996) apply partition technique to cluster similar trend and then on the basis of similar clusters identify similar characteristic, and ultimately, the products in each cluster have similar behavior in terms of sales. Sun et al. (2008) and Xia et al. (2012) employ extreme learning machine techniques for the goal sales forecasting in fashion retailing. Moreover, to make a prediction Bayesian techniques have been applied (Yelland and Dong 2014). Specifically, Ferreira et al. (2016) consider an online retailer and develops a prediction model. To do so, they use the following input factors: relative price of competing styles, discount, number of competing styles, brand manufacturer's suggested retail price index, color popularity, and finally price. Apart from forecasting on demand, there are researches focused on trend forecasting for the case of color (Gu and Liu 2010; Choi et al. 2014; Linton 1994; Diane and Cassidy 2009). In particular, Stansfield and Allan Whitfield (2005) use information in past on the use of prediction for future. Worth mentioning that in Liu et al. (2013), Choi et al. (2014) a survey on the existing approaches w.r.t. the sales forecasting especially related to *artificial intelligence* is conveyed.

It would be worthwhile to mention that sales forecasting is mainly counted as a regression problem and mathematically is tractable by applying regression techniques. Fallah Tehrani and Ahrens (2016b) cluster the similar data points, and on each cluster, a regression model is conducted. Dai et al. (2015), Lu (2014) apply state-of-the-art support vector machine regression (SRM) for the goal sales forecasting. Beyond modeling dependencies linearly, decision tree regression is able to capture nonlinear dependencies. In Atanackov and Boylan (2011), it has been proposed to use decision tree regression for forecasting trended demand. In addition, Bala (2010) deals with demand forecasting in retail sale. Kumar (2013) employs decision tree for wetter forecasting. Ulvila (1985) conveys a survey on decision trees underlying forecasting.

The terminology prediction is a widely used term in machine learning; however, this article refers to make a class prediction as well as the demand forecast of fashion products in the market.

3 Ordinal Logistic Regression (OLR)

Let us assume there are j classes, namely $\mathcal{Y} = \{y_1, \ldots, y_J\}$, where in addition they are ordered, i.e., $y_1 \prec y_2 \prec \cdots \prec y_J$. In an ordinal class, classification scenario is given n data points, which are supposed to be an *i.i.d* (independent and identically distributed):

$$\mathcal{D} = \left\{ (\boldsymbol{x}_i, y_i) \right\}_{i=1}^{n} \subset \mathbb{R}^m \times \mathcal{Y},$$

the goal is to learn a classifier $C : \mathbb{R}^m \to \mathcal{Y}$, which minimizes the following risk function under probability distribution $\mathbf{P}_{XY}$:

$$R(C) = \int_{X \times Y} l(C(\boldsymbol{x}), y) \, d \, \mathbf{P}_{XY}(\boldsymbol{x}, y) \,,$$

where $l(\cdot)$ is a loss function (error function). Since we are dealing with ordinal responses, one natural way to take into account the ordering is to consider the so-called cut-points (Bender and Grouven 1997). In fact, for each class y_l, one can assume a cut-point as follows:

- subclasses of y_l
- superclasses of y_l

The conditional probability that the class y observed for an instance $\boldsymbol{x}$ is at least y_l is termed as follows:

$$\pi_l(\boldsymbol{x}) := \mathbf{P}(y > y_l \mid \boldsymbol{x}) \,,$$

and consequently, the conditional probability that the class y observed for an instance $\boldsymbol{x}$ is at most y_l is equal to:

$$1 - \pi_l(\boldsymbol{x}) = \mathbf{P}(y \leq y_l \mid \boldsymbol{x}) \,,$$

leads to "binarize" the original problem keeping the order w.r.t the responses. We shall recall that if an instance $\boldsymbol{x}$ belongs to class y_l, then by the definition of ordinal classes the instance $\boldsymbol{x}$ belongs to all subclasses of y_l. This contribution is the core motivation to define cumulative *logit* as follows:

$$\log \left(\frac{\pi_l(\boldsymbol{x})}{1 - \pi_l(\boldsymbol{x})} \right) = \hat{w}_l + \sum_{i=1}^{m} w_i x_i \,.$$

Here, $\hat{w}_l$ is the intercept and $\{w_i\}_{i=1}^{m}$ are the corresponding weights. Roughly speaking, *logit* compares the probability estimation for classes that contain instance $\boldsymbol{x}$ to other classes. Note that in this case the slope for all classes is same, whereas only the intercept (bias) is changed. A simple calculation yields:

$$\frac{\pi_l(\boldsymbol{x})}{1 - \pi_l(\boldsymbol{x})} = \exp(\hat{w}_l) \exp\left(\sum_{i=1}^{m} w_i x_i \right) \,.$$

Ultimately, the probability that the class y is at least y_l is equal to:

$$\pi_l(\boldsymbol{x}) = \frac{\exp\left(\sum_{i=1}^{m} w_i x_i + \hat{w}_l \right)}{1 + \exp\left(\sum_{i=1}^{m} w_i x_i + \hat{w}_l \right)} \,, \tag{1}$$

and then, the probability that the class of an instance x is equal to $y = y_l$ is defined as follows:

$$P(y = y_l \mid x) = P(y > y_{l-1} \mid x) - P(y > y_l \mid x) = \pi_l(x) - \pi_{l-1}(x) \ ,$$

where $\pi_0(x) = 1$ and $\pi_J(x) = 0$. Since $\pi_l(\cdot)$ is increasing in l, this fact implies that for $0 \leq l < l+1 \leq J$, $\hat{w}_l < \hat{w}_{l+1}$, which imposes that

$$\hat{w}_1 \leq \cdots \leq \hat{w}_J \ . \tag{2}$$

For estimating the parameter from data, the likelihood function given n instances is termed as follows:

$$\mathcal{L}\left((\boldsymbol{w}, \hat{\boldsymbol{w}}) \ \Big| \ \left\{ (x_i, y_i) \right\}_{i=1}^{n} \right) = \prod_{i=1}^{n} \prod_{j=1}^{J} P\left(Y = j \ \Big| \ x_i \right)^{\mathbb{I}(y_i = j)} \ .$$

Accordingly, the log likelihood function is equal to:

$$\log \mathcal{L}\left((\boldsymbol{w}, \hat{\boldsymbol{w}}) \ \Big| \ \left\{ (x_i, y_i) \right\}_{i=1}^{n} \right) = \sum_{i=1}^{n} \sum_{j=1}^{J} \mathbb{I}(y_i = l) \log \left(\pi_j(x_i) - \pi_{j-1}(x_i) \right) \ .$$

Given a training data, then the optimal parameters are achieved by maximizing the (log) likelihood function, which in this case due to the constraints imposed by (2) the optimization is done by a constrained optimization setting.

3.1 Evaluation

Once the model has been trained, evaluation of an unseen instance x^* is carried out in the following way, considering:

$$\left(P(y_1 \mid x^*), \ldots, P(y_J \mid x^*) \right) \in [0, 1]^{\mathcal{Y}} \ ,$$

the vector of conditional probability, the prediction given an instance x^* is accomplished in the mode, namely:

$$\hat{y} = \underset{y_j \in \mathcal{Y}}{\operatorname{argmax}} \left\{ P\left(y_j \mid x^* \right) \right\} \ . \tag{3}$$

4 Regression Trees

In this section, we introduce a top-down greedy approach called Iterative Dichotomiser 3 (ID3) by Quinlan (1986), Vasudevan (2014). Roughly speaking, any decision tree attempts to partition the input space into several proper sub-input spaces, which elements in each sub-input space have a similar output characteristic. During partitioning, decision tree conducts a model on each sub-input space, and furthermore, algorithm develops incrementally binary tree by adding two edges into each node. As it will be clarified, through partitioning, indeed, decision trees build nonlinear decision boundaries, and henceforth, it can model as well as attribute inter-action. Before introducing the algorithm, we shall introduce the concept of entropy. From a statistical point of view, uncertainty quantification in data (S) is calculated by entropy:

$$H(S) = - \sum_{x \in X} p(x) \log_2 p(x) \ ,$$

where X is the set of classes in S and $p(x)$ determines the proportion of the number of elements in class x to the number of elements in S.

In the case of ID3, a decision tree is composed of several nodes starting from root node. Given a set of attributes, the root note is assigned to an attribute which has the smallest entropy (largest information gain). Once the root node has been identified, the algorithm extends the tree by adding the next attribute which has the smallest entropy among of rest attributes. The algorithm continues to recurse on attributes which never met so far. The algorithm is terminated then on each node if either every element in the subset belongs to the same class, or there are no more attributes or the node does not meet any training example. In this article, specifically, we deal with regression trees. A regression tree is analogous to an ordinal classification tree; namely, in this case, the output values are considered ordered. While in classification scenario output contains several classes, in the case of regression output may contain any real value. For regression case, however, instances in a sub-input space are assigned through output-averaged values from the same subspace. More concretely, in the regression case similarly algorithm begins with root node. Given a set of attributes, the root note is assigned to an attribute x_j which minimizes the deviation of a numerical sample. In the regression case, we apply the following summand:

$$\sum_{i|x_j \leq s} (y_i - \hat{c}_1)^2 + \sum_{i|x_j > s} (y_i - \hat{c}_2)^2 \ ,$$

where $\hat{c}_i = \frac{\sum_i y_i \mathbb{I}(x \in R_m)}{\sum_i \mathbb{I}(x \in R_m)}$; R_m is the rectangle (subspace) partitioned by thresholding on attributes. The algorithm consequently extends the tree by performing searches on the remaining attributes and chooses an attribute which minimizes the above summation and again attaches the node to the tree. Thus far, the fitting process was explained, and henceforth, the tree is capable to make forecast for unseen data points.

Fig. 1 Illustration of a
regression tree

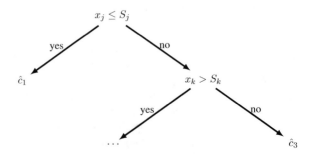

To conduct a reasonable decision on ordering, the decision maker requires a meaningful interpretation. A decision maker is not only interested in forecast values, but also fully interested in how he can modify the ordering through arranging order quantity. In many cases, companies would not follow exactly what the forecast machine offers, rather considering forecast values they are interested to adapt these values on ordering. This sounds reverse engineering; given sales values, what would be values of input factors? The common pitfalls for most of ML methods, except a few of them, are that this task is not manageable. For instance, for kernel machine regression the parameters α are not understandable, and hence, any change can cause a very flawed conclusion. In this regard, the decision tree offers this opportunity, and any binary decision tree can be rewritten as several rules. Of course, this property is quite desirable for any decision maker to fit commercial diplomacy on the ordering procedure. The tree in Fig. 1 can be written as several rules as follows:

```
IF xj > Sj THEN output = ĉ1 .
IF xj ≤ Sj AND xk > Sk THEN output = ĉ3 .
IF xj ≤ Sj AND xk ≤ Sk THEN output = ... .
```

Indeed, any branch in regression tree presents a rule, and hence, a whole tree can be reformulated as several rules. A plausible observation here is that a decision maker can modify the ordering process by rearranging the values on decision boundaries.

5 Algorithm

As mentioned earlier, identifying trendy products can cause higher gain in terms of performance, due to the fact that a wrong prediction for trendy products declines the quality of prediction drastically. Accounting for the fact that the amount of sales can be classified as "substantial," "middle," and "inconsiderable," the problem of estimating the trendy products boils down to identifying the classes of new products. Let us assume that the dataset $\mathcal{D} = \left\{ (x_i, y_i) \right\}_{i=1}^{n} \subset \mathbb{R}^{m+1}$ is given and moreover assumes $\left\{ y_i \right\}_{i=1}^{n}$ are the regression outputs (number of sales). Since the ordinal class classification setting receives the classified instances, the regression data must be converted to a labeled data. In this regard, it is needed to find two thresholds, as lower and upper bounds; the instances which are having an output smaller than lower bound belong to the class inconsiderable; the instances which are having an output higher than upper bound belong to the class substantial and respectively the instances which are having an output in between belong to the class middle. Since we are dealing with discrete values (number of sales), it is possible to group the output into three groups. In order to establish groups, we conduct a 3-means clustering solely on output (we indeed conduct a 3-means clustering using one-dimensional data). Now let us consider the instances are labeled by three clusters C_1, C_2, and C_3. The lower bound is determined as follows:

$$LB \leftarrow \min \left\{ \max_{(x,y) \in C_1} \{y\}, \max_{(x,y) \in C_2} \{y\}, \max_{(x,y) \in C_3} \{y\} \right\}. \tag{4}$$

Similarly, the upper bound is determined as follows:

$$UB \leftarrow \max \left\{ \min_{(x,y) \in C_1} \{y\}, \min_{(x,y) \in C_2} \{y\}, \min_{(x,y) \in C_3} \{y\} \right\}. \tag{5}$$

This categorization indeed classifies the instances without any conflict. Let us suppose that the output has been already clustered. Then, the learning problem can be tackled by the ordinal class classification scenario, where, here, there exists three ordinal classes. Henceforth, we call these classes as *substantial, middle* and *inconsiderable*. Formally, the classes are defined as follows:

$$\text{Substantial} := \left\{ (x, y) \in \mathcal{D} \mid y \geq UB \right\},$$

$$\text{Middle} := \left\{ (x, y) \in \mathcal{D} \mid LB < y < UB \right\},$$

$$\text{Inconsiderable} := \left\{ (x, y) \in \mathcal{D} \mid y \leq LB \right\}.$$

In the following, we discuss in greater detail the way to model these classes.

- **Classification**: Supposing three ordinal classes, namely "substantial" (S), "inconsiderable" (I), and "middle" (M), allows us to handle the problem in an ordinal class classification setting. As mentioned earlier, in the realm of forecasting the companies are not interested solely in an accurate prediction in terms of regression forecast. Rather, they require to identify whether a new product can be categorized as substantial. To this end, we suppose three ordinal classes, derived from number of sales in past. Then, the products assigned by corresponding classes are used to learn a proper model. More formally, in this case the likelihood function is formalized as follows:

$$\mathcal{L}\left(\boldsymbol{w}, \{\hat{w}_j\}_{j=1}^J \ \middle| \ \left\{product_i\right\}_{i=1}^n\right) = \prod_{i=1}^n \prod_{l \in \{S,M,I\}} \mathbf{P}\left(y_i = l \ \middle| \ product_i\right)^{\mathbb{I}(y_i=l)},$$

(6)

where S, I, and M stand for the substantial, the inconsiderable, and the middle, respectively. Finally, each product is associated with its class (y_i), and $\mathbb{I}(\cdot)$ stands for the indicator function. Supposing that the product i-th is given, then the probability distribution is calculated as follows:

$$\mathbf{P}\left(y_i = l \ \middle| \ product_i\right) = \pi_l\left(product_i\right) - \pi_{l-1}\left(product_i\right),$$

that is, the probability is that the $product_i$ belongs to the class l (see Fig. 2). For the cumulative distribution function ($\pi_l(\cdot)$), we consider the same setting in Eq. (1). Applying classical methods w.r.t. the maximum likelihood estimation, the goal is to find a vector $\boldsymbol{w}$ and intercepts $\{\hat{w}_j\}_{j=1}^J$, which maximize the likelihood function (maximize the likelihood of observed data points).

- **Regression**: A well-known drawback of decision trees is their sensitivity to perturbations in input space; i.e., any slight change in magnitude of input variables can effect on the response dramatically. At one extreme, a trendy product due to a minor change can be identified as a non-trendy product, which leads to a poor performance. To avoid this inconsistency, we partition the output space into several partitions, expecting that performing a regression tree on each partition reduces the sensitivity effect; the crux idea behind is that the products in each partition behave in a similar manner (in terms of output). Once the classes have been determined, for the instances, belonging to the same class a regression tree is fitted, however this time the regression response is considered, too. More concretely, let us assume that we are given three ordinal classes: substantial, middle, and inconsiderable with the ordinal logistic regression model. Basically, for each sub-dataset labeled by the same class, a regression tree model is fitted, and hence, there are three regression tree models. Henceforth, we call them $RT_S(\cdot)$, $RT_I(\cdot)$ and $RT_M(\cdot)$. By assuming that the probability that instance $\boldsymbol{x}$ belongs to the classes S, I, and M are dedicated by $\mathbf{P}(S \mid \boldsymbol{x})$, $\mathbf{P}(I \mid \boldsymbol{x})$, and $\mathbf{P}(M \mid \boldsymbol{x})$, and also, the fitted regression models are given by $RT_S(\cdot)$, $RT_I(\cdot)$, and $RT_M(\cdot)$, respectively, the core idea is to use the probabilistic information for the goal regression as follows:

$$\mathcal{P}_{\text{model}}(\boldsymbol{x}) = \mathbf{P}(S \mid \boldsymbol{x}) \cdot RT_S(\boldsymbol{x}) + \mathbf{P}(M \mid \boldsymbol{x}) \cdot RT_M(\boldsymbol{x}) + \mathbf{P}(I \mid \boldsymbol{x}) \cdot RT_I(\boldsymbol{x}).$$

(7)

The model serves several advantages: Firstly, it is coherent with the probabilistic nature of the ordinal logistic regression, which in fact can be seen as an advantage. Worth mentioning, the probabilistic models like (ordinal) logistic regression are robust to noises and outliers. Secondly, this representation tackles the learning problem in a probabilistic framework, that is, the probability that instance $\boldsymbol{x}$ belongs to class l. Note that it may be possible that an instance is *miss-classified*, thanks to a numerical problem or lack of information (one can imagine when all

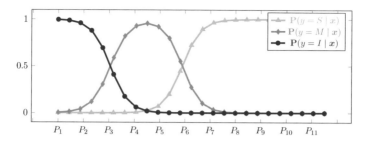

Fig. 2 Illustration of the conditional probability distribution $\left(\mathbf{P}\left(y=l\mid x\right)\right)$ for three ordinal classes considering several fashion products

three probabilities are close to 0.33, it is hard to verify a true class). Thereby, it prevents as much as possible this inconsistency, by assuming positive weights (probability) for each regression estimator. Thirdly, the probability of each class given a product can be interpreted as a degree of membership w.r.t. each class. The three classes can be considered as three features contributing the member-ship, which can be interpreted as the importance of each feature with respect to the instance x. Seen from this view, it has the same setting similar to the weighted mean. Lastly and specifically, this setting filters outliers as it categorizes the output in advance. Note that the expression in (1) can be reformulated as follows:

$$\pi_l(x) = \frac{\exp\left(\gamma\left(\sum_{i=1}^{m} w_i^* x_i + w^*_l\right)\right)}{1 + \exp\left(\gamma\left(\sum_{i=1}^{m} w_i^* x_i + w^*_l\right)\right)} \, ,$$

where $\gamma > 0$, $\mid w_i^* \mid \leq 1$, and $\mid w_l^* \mid \leq 1$. The parameter γ indicates the precision of the model; i.e., the higher the parameter γ, the more accurate the model. In extreme case, when $\gamma \rightarrow \infty$ the sigmoid function becomes step function, see Fig. 3. For the case of large γ, the class of instances are assigned precisely, and hence, there is one regression model required. In this case for other classes, $\mathbf{P}(C \mid x) \rightarrow 0$, and therefore,

$$\mathcal{P}_{\mathrm{model}}(x) \approx \mathbf{P}(C^* \mid x) \cdot R_{C^*}(x) \, ,$$

where C^* is the true class. In the case of poor γ, there is an uncertainty in model. To overcome this, the model considers positive weights (probability) for the regression estimators, which in fact reflects the importance of each class w.r.t. the instance concurrently. Roughly speaking, the weights can be thought of as several switches, and in the case of ambiguity, they are switched on, otherwise all except one are off.

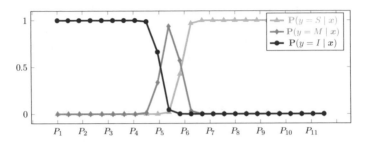

Fig. 3 Illustration of the conditional probability distribution $\left(\mathbf{P}\left(y = l \mid x\right)\right)$ for three ordinal classes considering several fashion products for a large value of γ

6   Experiments

We tested the feasibility of our approach to classify and make an accurate prediction on a real dataset from an apparel industry, which is taken from 2011–2014.

6.1   Datasets

We performed a case study in the fashion sector, where there is the potential to monitor the feedback of customers over time. The dataset is acquired from an apparel retailer in Bavaria (Germany) and in particular the sales information derived from apparel catalogues. Before the begin of a new clothing season, the retailer sends catalogues to its customers from which its customers can order products. Simultaneously or even earlier, the retailer orders fashion products from other upstream retailers, most of which are located in Asia. Therefore, an accurate sales-ordering is sought and leads to decrease in unnecessary expenditures. As mentioned earlier, the most advertised fashion products are distinctive from year to year. Seen from this view, the estimator is required to recognize the proper patterns from the sales history.

The dataset in total contains 2552 products, where this value is reduced to 150 different product categories. Each product is associated with the following attributes: *color, size of photograph, price, and the number of colors for a product* with *the number of sales* as the output. A summary of datasets is illustrated in Table 1. The size of photograph, the price, the number of colors for the product and the output are real-valued features (quantitative features), whereas color is a qualitative feature. In order to convert the qualitative feature (color) to a quantitative feature, we clustered the colors into 4–6 groups derived by their similarities. This procedure has an obvious advantage; namely, it permits the consideration of several trend colors at the same time. As is clear, there is a spectrum of the trend color and usually would not be assigned to a specific color. In addition, similar colors incorporate into nearly the same way according to the response, in the sense that they have the same weights,

Table 1 Description of datasets over several years (2011–2014)

Description of datasets				
Attributes	Color	Size of photograph	Price	Number of colors
Type of attribute	Qualitative	Quantitative	Quantitative	Quantitative
M^a (μ)	–	20.09	110.07	3.31
Median	–	9.20	89.95	2
SD^b	–	26.15	65.94	3.16
Size of datasets				
Years	2011	2012	2013	2014
Number of data	621	662	699	570
Number of sales (output)				
M (μ)	292.69	283.10	291.19	264.75
Median	202	178	194	176
SD	285.05	286.96	297.36	307.19

[a]M stands for the mean
[b]SD stands for the standard deviation

which in this setting obviously is considered. In sequel, only the categories which have more than 20 training are considered. This filtering indeed reduces the effect of overfitting.

6.2 Experimental Setup and Evaluation

We conducted two types of experiments, namely for the classification setting and for the regression setting. For both settings, we split the data into two parts: one part for training the model and the rest for the evaluation. To this end, we made a cross-validation on years, namely three years for training process, and evaluation is carried out on the rest. In order to consider all possible combinations over years, this procedure is repeated for all four years. In other words, we consider all "3-combinations" from the set $\{2011, 2012, 2013, 2014\}$ for fitting the model, what remains is applied for the evaluation phase. Moreover, we conduct the experiments regarding the product category in a sense that each experiment considers solely the products in the same category and finally, the results are averaged.

Let us assume that the training data have been already chosen; the two thresholds, lower bound and upper bound for the discretization the regression output into the three ordinal classes are adjusted by the settings in (4) and (5). These bounds allow for converting the regression outputs to the ordinal labels. The input attributes in the training set then are assigned to the ordinal classes, and hereupon, the dataset is a suitable dataset for the goal ordinal classification, in particular for the setting in (6).

Once the model has been trained, it can be used for the goal prediction; namely, given a testing data, the prediction is then performed by (3). The two bounds decompose the training data into three parts, namely the part which has output values smaller than LB, the part which has output values between LB and UB, and finally the part which has output values larger than UB. For the regression scenario, for each part a regression tree is fitted, and in the end, the three regression trees are weighted by the probabilities derived from the classification part. Henceforth, the model is used for the goal prediction.

In addition, as baseline model, we assume common regression tree, which is trained on the whole training data without assuming any data separation. Indeed, we fit regression trees on regression data, and since the regression results can be converted into ordinal classes (by thresholding on regression output), the extracted results also can be used immediately for the goal ordinal class classification prediction. This comparison is necessary to illustrate that our modification despite of more computational complexity achieves a significant gain. To justify the performance comparison, two types of loss functions have been applied for the two settings (classification and regression), namely:

I. The classification accuracy is measured by normalized absolute error:

$$\frac{1}{2n} \sum_{i=1}^{n} | y_i - \hat{y}_i | \, ,$$

where $\hat{y}_i$ is a predicted class given product ith. In this setting factor, 2 is considered to normalize the results between $[0,1]$.

II. The regression accuracy is measured in terms of mean absolute percentage error (MAPE):

$$\frac{1}{n} \sum_{i=1}^{n} \left| \frac{y_i - \hat{y}_i}{y_i} \right| \, ,$$

where $\hat{y}_i$ is a predicted response value, whereas y_i is the actual response of the instance x_i.

6.3 Results

In Tables 2 and 3, the results for the classification and the regression are depicted. In both tables, the years are associated with the testing data and **LR, OLR, RT** are linear regression, ordinal logistic regression, and regression tree, respectively. Here, the prefix modified is referred to our proposal, namely the regression trees by the probabilities derived from the ordinal logistic regression. Finally, the errors are averaged for both loss functions. At first glance, it may seem that for both settings the error results should be correlated. However, while in the classification scenario **OLR**

Table 2 The averaged results for whole products in terms of normalized classification error

Methods	2011	2012	2013	2014
LR	0.38	0.39	0.38	0.36
OLR	0.41	0.35	0.36	0.33
RT	0.38	0.39	0.30	0.39

Table 3 The averaged results for whole products in terms of mean absolute percentage error

Methods	2011	2012	2013	2014
Mean	2.22	2.88	2.74	2.56
LR	1.31	1.79	1.64	1.66
Modified LR	1.15	1.59	1.40	1.41
RT	1.29	1.80	1.40	1.31
Modified RT	1.09	1.28	1.37	1.15

delivers the best performance, in the regression scenario **RT** (modified versions) is indeed superior. The reason first of all is that the ordinal logistic regression exploits the so-called prior knowledge in the sense that the discretized data contain more information regarding ordinal classes. However, the other models do not suppose this discretization. Notably, such discretization is more in the benefit of linear models because it reduces the effect of the obstacles (outliers and noises). As stated at the beginning, the datasets are extracted from a catalogue. As might be expected for the most cases, the colors are the most important factor from an ORL point of view, which depends on products, and clustering its importance varies. The next most important features tended to be the number of colors offered for a product and the size of photograph.

In the regression scenario, the **RT** takes advantage of its nonlinear flexibility. Another interesting observation is the improvement of the modified version over its competitors in the regression case. This again confirms that incorporating the prior knowledge in the regression trees increases the performance. Apart from the improvement in terms of MAPE, it is worthwhile to mention that the results of MAPE show high errors. The common RT without any separation on data indicated that for best sellers the size of photograph plays an important role, while for worst sellers the colors and price contribute, additionally.

We conducted another experiment which considers a naïve approach; namely, we took the mean of number of sales in training data and applied it to the prediction value. The results are depicted in Table 3; the results show a very high error, specifically, compared to our achievement. Roughly speaking, the error values of mean approach may imply that how challenging the forecasting problem is. Note that in our setting the mean value concerns only the number of sales of similar products in a group and hence is a valid forecast value.

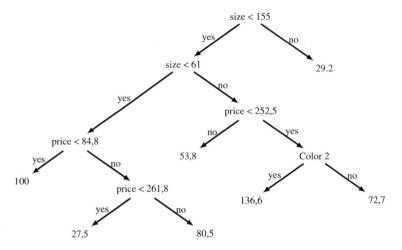

Fig. 4 Illustration of the decision tree for whole data

6.4   Tree Illustration

As it is mentioned at the beginning of this paper, a well-known advantage of decision trees is comprehensibility. In this section, we illustrate a concrete case: For this case study, we took a product from dataset, namely bag. We conducted two types of models, RT and modified RT. In both cases, we plotted the fitted regression trees, and also, we computed the performance.

The common regression tree fitted on the whole dataset is depicted in Fig. 4; it contains size of photograph, price, and color2. In each node, if constraint is satisfied left edge is followed, otherwise right edge and finally the decision is made at the end of tree. A particular disadvantage of such tree, however, is their sensitivity to perturbations in input values; i.e., any slight change near the node values may cause a large difference in terms of output prediction. For instance, for an area value around 155 there is a high risk to choose a wrong edge. Another drawback of decision tree is that they can be easily overfitted, which can occur in the lack of data—Unfortunately, there is no regularization option for RT. These drawbacks motivate us to prevent these problems by applying more simple trees, which have a less number of nodes and edges. To this end, in fact, through discretization the training examples have a more homogeneous structure and hopefully trees require less number of nodes. To illustrate this, we plotted also the fitted trees for modified version; the trees are illustrated in Figs. 5, 6, and 7. As can be seen, applying more homogeneous data leads to less complex trees. Figure 5 demonstrates that for inconsiderable part the tree only contains one node, namely price; price is higher than 168.3, the output is 8.5, otherwise is 20. While for inconsiderable case the decision tree is composed of only one node, for middle case the decision tree is more complex, due to the fact that for middle part more data points exist and perhaps more inhomogeneity in this

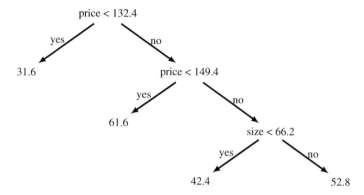

Fig. 5 Illustration of the decision tree for the inconsiderable part

Fig. 6 Illustration of the decision tree for the middle part

Fig. 7 Illustration of the decision tree for the substantial part

part. For substantial part again, the tree has one node, namely price, perhaps due to the less number of training examples. In sequel, the performance of common RT and modified RT is reported as follows: For the case of common RT, the MAPE is 0.7614, whereas for the modified RT, it is reduced to 0.7028.

7 Concluding Remarks

In continuation of our previous work w.r.t. predictive models Fallah Tehrani and Ahrens (2016a), in this study we applied a nonlinear interpretable model, namely regression tree. The combined model underlying ordinal logistic regression incorporates with the prior knowledge and in addition is more robust to the obstacles. Moreover, we showed a case study in which the illustration of regression tree and the benefit of embedding trees in a probabilistic framework have been illustrated.

This, in fact, does not only contribute to gain but also can serve a simpler interpretation for any adjustment in future. First experimental studies have shown that this method compares quite favorably with linear regression and other competitors. We are convinced of its high potential in the combined model, and in future, we are targeting exploiting the benefits of our proposal on the use of more real applications, in particular in terms of any tuning of sales volume. Worth mentioning, typically apparel industries have every year a new sell strategy due to budget, human resources, facilities, and so on. To adopt new plans on existing information from previous year, they have to retune input factors. The interpretable models offer this opportunity to adopt a new plan on sales history.

References

Alon I, Qi M, Sadowski RJ (2001) Forecasting aggregate retail sales: a comparison of artificial neural networks and traditional methods. J Retail Consum Serv 8(3):147–156

Atanackov NN, Boylan JE (2011) Service parts management: demand forecasting and inventory control, chapter decision trees for forecasting trended demand. Springer, London, pp 53–87

Bala PK (2010) Decision tree based demand forecasts for improving inventory performance. In: 2010 IEEE international conference on industrial engineering and engineering management (IEEM). Dec 2010, pp 1926–1930

Bender R, Grouven U (1997) Ordinal logistic regression in medical research. J R Coll Phys Lond 31(5):546–51

Choi TM, Hui CL, Yu Y (eds) (2014) Intelligent fashion forecasting systems: models and applications. Springer Publishing Company, Incorporated

Dai W, Chuang YY, Lu CJ (2015) A clustering-based sales forecasting scheme using support vector regression for computer server. Proced Manuf 2:82–86. In: 2015 2nd international materials, industrial, and manufacturing engineering conference, MIMEC2015, 4–6 February 2015, Bali, Indonesia

Diane T, Cassidy T (2009) Colour forecasting. Wiley-Blackwell

Fallah Tehrani A, Ahrens D (2016a) Enhanced predictive models for purchasing in the fashion field by using kernel machine regression equipped with ordinal logistic regression. J Retail Consum Serv 32:131–138

Fallah Tehrani A, Ahrens D (2016b) Improved forecasting and purchasing of fashion products based on the use of big data techniques. Springer Fachmedien Wiesbaden, Wiesbaden, pp 293–312

Ferreira K, Alex Lee BH, Simchi-Levi D (2016) Analytics for an online retailer: demand forecasting and price optimization. Manuf Serv Oper Manag 18(1):69–88

Frank C, Garg A, Sztandera L, Raheja A (2003) Forecasting women's apparel sales using mathematical modeling. Int J Cloth Sci Technol 15(2):107–125

Gu W, Liu X (2010) Computer-assisted color database for trend forecasting. In: 2010 international conference on computational intelligence and software engineering (CiSE), December 2010. IEEE, pp 1–4

Happiette M, Rabenasolo B, Boussu F (1996) Sales partition for forecasting into textile distribution network. In: 1996 IEEE international conference on systems, man, and cybernetics, October 1996, vol 4, pp 2868–2873

Huang H, Qiurui L (2017) Intelligent retail forecasting system for new clothing products considering stock-out. Fibres Text East Eur 25(1)

Kit Wong W, Guo Z (2010) A hybrid intelligent model for medium-term sales forecasting in fashion retail supply chains using extreme learning machine and harmony search algorithm. Int J Prod Econ 128(2):614–624

Kumar R (2013) Decision tree for the weather forecasting. Int J Comput Appl 76(2):31–34

Kuo R, Wu P, Wang C (2002) An intelligent sales forecasting system through integration of artificial neural networks and fuzzy neural networks with fuzzy weight elimination. Neural Netw 15(7):909–925

Linton H (1994) Color forecasting: a survey of international color marketing. Van Nostrand Reinhold

Liu N, Ren S, Choi TM, Hui CL, Ng SF (2013) Sales forecasting for fashion retailing service industry: a review. Math Probl Eng

Lu C (2014) Sales forecasting of computer products based on variable selection scheme and support vector regression. Neurocomputing 128:491–499

Quinlan JR (1986) Induction of decision trees. Mach Learn 1(1):81–106

Rodrigues EM, Figueiredo MC (2013) Forecasting demand in the clothing industry. In: XI Congreso Galego de Estatística e Investigación de Operacións

Stansfield J, Allan Whitfield TW (2005) Can future colour trends be predicted on the basis of past colour trends?: an empirical investigation. Color Res Appl 30(3):235–242

Sun ZL, Choi TM, Au KF, Yu Y (2008) Sales forecasting using extreme learning machine with applications in fashion retailing. Decis Support Syst 46(1):411–419

Thomassey S, Happiette M (2007) A neural clustering and classification system for sales forecasting of new apparel items. Appl Soft Comput 7(4):1177–1187

Thomassey S, Happiette M, Castelain JM (2002a) Textile items classification for sales forecasting. In: Proceeding 14th european simulation symposium (ESS)

Thomassey S, Happiette M, Dewaele N, Castelain JM (2002b) A short and mean term forecasting system adapted to textile items' sales. J Text Inst 93(3):95–104

Ulvila JW (1985) Decision trees for forecasting. J Forecast 4(4):377–385

Vasudevan P (2014) Iterative dichotomiser-3 algorithm in data mining applied to diabetes database. J Comput Sci 10(7):1151–1155

Xia M, Zhang Y, Weng L, Ye X (2012) Fashion retailing forecasting based on extreme learning machine with adaptive metrics of inputs. Knowl Based Syst 36:253–259

Yelland PM, Dong X (2014) Forecasting demand for fashion goods: a hierarchical bayesian approach. In: Choi TM, Hui CL, Yu Y (eds) Intelligent fashion forecasting systems: models and applications. Springer, Berlin, Heidelberg, pp 71–94

A Data Mining-Based Framework for Multi-item Markdown Optimization

Ayhan Demiriz

Abstract Markdown decisions in retailing are made based on the demand forecasts which may or may not be accurate in the first place. In this chapter, we propose a framework for forecasting weekly demands of retail items via linear regression models within multi-item groups that incorporate both positive and negative item associations. We then utilize dynamic pricing models to optimize markdown decisions based on the forecasts within multi-item groups. Grouping items can be considered as a form of variable selection to prevent the overfitting in prediction models. We report regression results from multi-item groupings besides results from single-item regression model on a real-world dataset provided by an apparel retailer. We then report markdown optimization results for the single items and multi-item groupings that multi-item forecasting models are built upon. The results show that the regression models provide better estimates within multi-item groups compared to the single-item model. Moreover, the overall revenues achieved in multi-item markdown optimization across all grouping schemes are higher than the total revenue yielded by single-item markdown optimization scheme.

1 Introduction

Retail decision-making processes involve activities such as pricing, allocation, and inventory management (Kumar and Patel 2010). As important inputs in decision processes, accurate forecasts can improve the retailers' profitability. Price markdowns in apparel industry are applied most of the time to the items with demands that are inaccurately forecasted in the first place. Sub-optimal decisions based on inaccurate forecasts eventually force retailers to eliminate the end-of-season inventories at salvage values. Therefore, accurate forecasts, especially in apparel retailing, are

A. Demiriz (✉)
Department of Industrial Engineering, Gebze Technical University, Kocaeli, Turkey
e-mail: ademiriz@gmail.com

© Springer Nature Singapore Pte Ltd. 2018
S. Thomassey and X. Zeng (eds.), *Artificial Intelligence for Fashion Industry in the Big Data Era*, Springer Series in Fashion Business,
https://doi.org/10.1007/978-981-13-0080-6_4

47

important for determining the initial inventory levels and the prices of the items as well as markdown pricing decisions during the retail seasons. We propose a framework in this work to determine optimal price markdowns within multi-item groups which utilizes multi-item demand functions based on various item (product) grouping schemes in which product associations are also considered.

Ever-increasing use of time series models in marketing science (Dekimpe and Hanssens 2000) opens a variety of tools to the decision makers' disposal. Well-known methods such as exponential smoothing (Gardner 2006) are the tools of the trade in forecasting for a variety of applications, e.g., revenue management in hospitality industry (Gardner 2006). Larger datasets in terms of number of variables and length of the time span in addition to determining aggregation granularity (both temporal and entity-wise) stand as some of the challenges for the decision makers. These challenges motivate for increasing the application of data mining in time series analyses (Dekimpe and Hanssens 2000; Fu 2011).

Apparel retailers need to overcome some analytical modeling challenges when it comes to forecasting. Each year thousands of new items, with relatively short life spans (around 6–12 weeks) and long lead-times (6–10 months), are introduced (Thomassey and Fiordaliso 2006). During the periods when two main collections (i.e., summer and winter collections) are introduced, thousands of items are replaced within a few weeks of the new season. Retailers face a forecasting problem that calls for the use of various explanatory variables to reflect the effects of weather, holidays, marketing actions, promotions, fashion, or economic environment (Thomassey and Fiordaliso 2006).

While the price of an item is an important predictor (explanatory) variable of that item's demand, it can also be used for predicting (forecasting) the demands of the related items. The term "related items" covers both complementary and substitute items that are linked with the pricing of that particular item. Two or more items are considered as complements (substitutes) if decreasing (raising) the price of one item leads to an increase in sales of another (Shocker et al. 2004). The relationships between item prices can be not only intra-categories (traditionally defined) but also inter-categories (Nijs et al. 2001). Moreover, quantifying the substitution effects based on only positive item associations (Vindevogel et al. 2005) may be problematic and misleading. Since both grouping items and determining complementary and substitute items are the essential part of the analytical studies in a retailing frame, finding the item associations and item grouping should be coupled in an analytically sound approach which is one of the principal aims of this study.

The main steps of the framework proposed in this work can be listed as follows: Using data mining for determining item associations to characterize the complementary and the substitution effects; identifying the groups of items that are related, complementary but not substitutes, and that have similar sales profiles (i.e., sales figures); then applying the forecasting models within these groups (Demiriz et al. 2010a); and finally implementing price markdown optimization within multi-item groups. When we consider substitute products within the same multi-item group, we may not be able to represent cannibalization effects properly in our price markdown models. This may further complicate the approach by including unforeseen factors. Therefore, it may be useful to keep substitute products out of multi-item groups. It may not be straightforward to think that negative associations could be used in place of substitutions but they may indicate assortment-based (Gürhan Kök and Fisher 2007), stock-out, and price substitutions not only within a single category but also across many categories (Demiriz et al. 2010b, 2011). The framework forecasts the weekly sales of retail items via multiple linear regression models that incorporate predictors that represent both positive and negative item associations. In addition to the pairwise item association constraints obtained through utilizing the transactional data (Demiriz et al. 2009, 2010b, 2011), our framework uses the item similarities based on weekly sales figures to group the similar items (Demiriz et al. 2010a). Grouping the items can be regarded as a form of variable selection in the regression models, which prevents overfitting compared to the case of utilizing the full dataset in the regression models. We propose three grouping schemes as follows: A heuristic grouping scheme is based on product associations, k-means clustering, and constrained k-means clustering. Once the demand functions are formed within a multi-item group by appropriate regression models, we can then utilize a deterministic optimization model to find the optimal price markdowns within the multi-item group. As benchmark alternatives to our framework, results based on single-item models are also reported later in the experimental study.

In short, our approach aggregates items according to both positive and negative association rules to forecast multi-item group demand, and then use a mathematical programming model to find optimal markdown strategy for those multi-item groups. By utilizing association mining and constrained clustering, an idea adapted from machine learning and artificial intelligence, we propose an intelligent support system to provide a concurrent solution to forecasting, optimization, and pricing problems.

The remainder of the chapter is organized in a way to follow our methodological approach, i.e., grouping the items first, then forecasting within multi-item groups and finally markdown price optimization: Sect. 2 introduces our methodology to group related items in terms of both complementary and substitution effects. Section 3 summarizes the related literature on the multi-item forecasting in the retail domain. Section 4 introduces deterministic multi-item markdown optimization model. In Sect. 5, empirical results are reported. Finally, Sect. 6 concludes this chapter. Some of the material shown in this chapter previously published in our earlier work in Demiriz (2015).

2 Grouping-Related Items

Our aim is to apply multivariate regression models within the item groups that are similar with respect to their weekly sales figures. The basic idea for grouping the items is that the attributes of similar items (e.g., price) may have better predictive properties for the focal item. Since an item can be placed into its category (within the product hierarchy) by some business purposes, a category naturally determines the related items. However, this type of grouping prevents modeling the possibilities

Fig. 1 Depiction of
traditional complementary
and substitution effects
(adapted from Shocker et al.
2004)

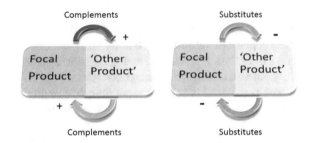

of inter-category relationships. Therefore, grouping based on category information
(i.e., item hierarchy) is not considered in this study.

The traditional complementary and substitution effects can be depicted as in Fig. 1.
For example, let us assume that our focal product is a v-neck sweater. A shirt is
certainly complementary to a v-neck sweater. On the other hand, a crewneck sweater
is a substitute to a v-neck sweater. Basically, a change in the price of the v-neck
sweaters can affect the overall sales of the shirts and crewneck sweaters based on
the relationship type. However, the nature of complementary and substitution effects
is not static and might change (Shocker et al. 2004). Therefore, some items that are
initially complementary may become substitutes for each other (Shocker et al. 2004)
over time. Nevertheless, the information on complementary and substitution effects
between items can naturally constitute a baseline to determine the related items.

In this chapter, we use two different types of similarity measures for determining
the related item groups. The first one depends on the item associations (both positive
and negative) found by association mining. The second one is based on the weekly
sales figures. Based on these two types of similarity measures, we propose three
different grouping schemes to determine the related item groups. The first grouping
scheme is a heuristic method based on the item associations. The second grouping
scheme is based on a clustering approach by using the k-means algorithm in which
the weekly sales figures are used for computing the similarity measure. The third
scheme is based on constrained clustering, in which both item associations and
the weekly sales figures are used for determining multi-item groups. These three
grouping schemes are introduced in this section. We assume that the associated
items can be considered as complementary and substitute.

Association mining is utilized in this study as a way of finding complementary and
substitute items. We follow the methodology used in our earlier work (Demiriz et al.
2010b, 2011) to find positive and negative associations that exist in our transactional
data. Particularly, negative associations are found by using indirect association mining (Tan et al. 2001). An example of negative association can be seen in dotted line
of Fig. 2. Assume there exist two positive associations: Shirt—V-Neck Sweater and
Shirt—Crewneck Sweater among Shirt, V-Neck and Crewneck Sweaters, since they
are frequent. Therefore we can conclude that an indirect association exists between
V-Neck and Crewneck Sweaters via Shirt (Tan et al. 2001). In the remainder of this
section, various grouping schemes are introduced to implement our approach.

Fig. 2 Depiction of indirect association

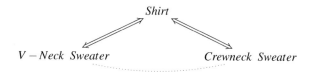

2.1 Associated Group Heuristic

Association mining is often promoted as an alternative way for deriving the item pro-motion strategies instead of traditional marketing approaches. However, it is reported in Vindevogel et al. (2005) that using item associations might be misleading in deter-mining the promotion strategies in light of some cross-price elasticity results that were found. The analyses in Vindevogel et al. (2005) are conducted by using transac-tional data from a European retailer (possibly not an apparel retailer). Approximately 2,700 pairwise item associations are found in the study. By computing cross-price elasticities, 1,112 associations are classified as complementary, 1,212 associations are classified as substitute, and 376 associations are classified as independent. There-fore, the authors in Vindevogel et al. (2005) rightfully claim that it is more meaningful to use cross-price elasticities for determining the item promotion strategies instead of only association mining results. This is due to the fact that all the item associations found would have been treated as complements if association mining results were used for determining the promotion strategies. However, some of them are indeed substitutions.

Nevertheless, the argument in Vindevogel et al. (2005) can be implicitly used to show that it is practical to use the association mining results for determining the related items, regardless of being either complementary or substitutes. In our approach, we employ the negative associations in addition to the positive ones. We devise the following heuristic method to determine the related item groups in the following steps:

Step 1: Create an item list by sorting all items by their total sales volumes in descend-ing order.

Step 2: Pick the item on the top of the list as the focal item.

Step 3: Select up to two positively associated items with the focal item from the list by searching the list from top to bottom (if any matches exist).

Step 4: Select up to two negatively associated items with the focal item from the list by searching the list from bottom to top (if any matches exist).

Step 5: Remove the focal item and the associated items from the list to form the associated items group.

Step 6: If there is no item in the list, then stop. Otherwise go to Step 2.

In order to depict associated group heuristic better, assume that we have the sales data given in Table 1. Notice that items are ordered by their total sales. Let us assume that the following positive associations are found to be frequent and are sorted in descending order of their frequencies (support counts): (AB, DB, BC, CF) and the

Table 1 Sample sales data

Item	Total sales
B	175
A	160
C	150
D	125
F	100

Fig. 3 Sample-associated group

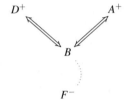

focal product is B. By applying the indirect association, one can easily conclude that items B and F are negatively associated via item C. Since there can be only up to two positively and negatively associated products with focal product, a group can be formed as in Fig. 3. Note that the types of associations with the item B are marked as superscript next to the items in Fig. 3. Item C is not part of this associated group as itemsets AB and DB have higher support count than itemset BC. Therefore, item C is omitted in Fig. 3.

The above heuristic may find the groups with up to five items including the focal item. However, the execution of this heuristic may result in some groups that have only the focal item. Nevertheless, this heuristic forms non-overlapping clusters that can be used for multivariate regression models.

2.2 k-Means Clustering

The natural way of determining the related items is to use the category information. However, a clustering algorithm can be practically used as well to find the related items (Kumar and Patel 2010). Actually this makes more sense, since a similarity measure is used for any clustering algorithm. As in Kumar and Patel (2010), weekly sales figures can be used to determine the similarity between items.

k-means is one of the most common clustering algorithms (Wu et al. 2008), and it can be formulated as a mathematical program for clustering m items over a period of t weeks in the following way. Given a dataset $\mathcal{X} = \{x_i\}_{i=1}^{m}$ of m points (items) in $\mathbb{R}^t$ and k desired clusters, the problem is to determine cluster centers $\mathcal{C}_1, \mathcal{C}_2, \ldots, \mathcal{C}_k$ in $\mathbb{R}^t$ such that the sum of the squared distance between each point x_i and its *nearest* cluster center $\mathcal{C}_l$ is minimized (Demiriz et al. 2018). Specifically:

$$\min_{\mathscr{C}_1,\ldots,\mathscr{C}_k} \sum_{i=1}^{m} \min_{l=1,\ldots,k} \left(\frac{1}{2} \|x_i - \mathscr{C}_l\|^2 \right) \tag{1}$$

By Bradley et al. (1997, Lemma 2.1), Eq. 1 is equivalent to the following problem where it is possible to remove the inner min operator in Eq. 1 by introducing "assignment" variables $Y_{i,l}$, $i = 1, \ldots, m$, $l = 1, \ldots, k$. Note that $Y_{i,l} = 1$ if data point x_i is closest to center $\mathscr{C}_l$ (i.e., belongs to cluster l) and zero otherwise.

$$\underset{\mathscr{C},Y}{\text{minimize}} \quad \sum_{i=1}^{m} \sum_{l=1}^{k} Y_{i,l} \cdot \left(\frac{1}{2} \|x_i - \mathscr{C}_l\|^2 \right)$$

$$\text{s.t.} \quad \sum_{l=1}^{k} Y_{i,l} = 1, \ i = 1, \ldots, m, \tag{2}$$

$$Y_{i,l} \geq 0, \ i = 1, \ldots, m, \ l = 1, \ldots, k.$$

Notice that m items can be grouped into k non-overlapping clusters by using t weekly sales figures to compute the similarity measure in k-means algorithm. The similarity measure in this case is the Euclidian distance, as seen in the objective function of Eq. 2. Alternatively, overlapping clusters can also be constructed without specifying the number of clusters, i.e., k (Hasan et al. 2011). However, it is more preferable to assign each item to a unique cluster from a business point of view.

2.3 Constrained Clustering

By adding new constraints to k-means formulation, we can easily define a constrained k-means clustering problem that incorporates item associations for finding the related item groups. The basic idea in this type of item grouping is that the groups are composed of positively associated items and the negatively associated items are placed into the different groups. In some cases, minimum cluster size constraints can also be put into effect to prevent the sub-optimal local solutions of k-means algorithm, e.g., single points forming some groups or even empty clusters. The positive associations (i.e., complementary relationships) can be represented by must-link constraints and the negative associations, i.e., substitution effects can be represented by cannot-link constraints.

There exists very efficient network algorithms to solve the problem defined in Eq. 2 (Demiriz et al. 2018). Must-link and cannot-link constraints, as well as the minimum cluster size constraints, can easily be introduced to the clustering problem defined in Eq. 2 as hard constraints, resulting in the following problem formulation:

$$\underset{\mathscr{C},Y}{\text{minimize}} \quad \sum_{i=1}^{m}\sum_{l=1}^{k} Y_{i,l} \cdot \left(\frac{1}{2}\|x_i - \mathscr{C}_l\|^2\right) \tag{3}$$

$$\text{s.t.} \quad \begin{aligned} & \sum_{l=1}^{k} Y_{i,l} = 1, \ i = 1,\dots,m, \\ & \sum_{i=1}^{m} Y_{i,l} \geq \tau_l, \ l = 1,\dots,k \\ & Y_{i,l} = Y_{j,l} \ i,j \in \{c_=(i,j)\} \ l = 1,\dots,k \\ & Y_{i,l} + Y_{j,l} \leq 1 \ i,j \in \{c_{\neq}(i,j)\} \ l = 1,\dots,k \\ & Y_{i,l} \geq 0, \ i = 1,\dots,m, \ l = 1,\dots,k. \end{aligned}$$

Here, the set $\{c_=(i,j)\}$ $i,j = 1,\dots,m$ represents the must-link constraints and the set $\{c_{\neq}(i,j)\}$ $i,j = 1,\dots,m$ represents cannot-link constraints. However, if must-link and cannot-link constraints are introduced as hard constraints as in the problem in Eq. 3, it is very likely that the underlying clustering problem will have no feasible solution at all for the large problems (depending on the parameter settings). This is not a desirable outcome of a clustering algorithm. The algorithm should be able to produce the cluster centers with as few constraint violations as possible. As in Lange et al. (2018), Lu and Leen (2018), Pensa et al. (2018), Lagrangian relaxation of the optimization problem defined in Eq. 3 can be formulated to overcome the infeasibility problem as follows.

$$\underset{\mathscr{C},Y}{\text{minimize}} \quad \begin{aligned} & \sum_{i=1}^{m}\sum_{l=1}^{k} Y_{i,l} \cdot \left(\frac{1}{2}\|x_i - \mathscr{C}_l\|^2\right) + \frac{\lambda^+}{2} \sum_{i,j\in\{c_=(i,j)\}|(Y_{i,l}\neq Y_{j,l}),l=1,\dots,k} 1 \\ & + \lambda^- \sum_{i,j\in\{c_{\neq}(i,j)\}|(Y_{i,l}+Y_{j,l}==2),l=1,\dots,k} 1 \end{aligned} \tag{4}$$

$$\text{s.t.} \quad \begin{aligned} & \sum_{l=1}^{k} Y_{i,l} = 1, \ i = 1,\dots,m, \\ & \sum_{i=1}^{m} Y_{i,l} \geq \tau_l, \ l = 1,\dots,k \\ & Y_{i,l} \geq 0, \ i = 1,\dots,m, \ l = 1,\dots,k. \end{aligned}$$

where λ^+ and λ^- are penalty parameters for the respective constraint violations in Lagrangian relaxation. The constraint violations are represented as a logical value where $Y_{i,l} \neq Y_{j,l}$ case is a violation of a must link and $Y_{i,l} + Y_{j,l} == 2$ case is a violation of cannot-link constraint considering the decision variable, $Y_{i,l}$, is a **0/1** variable. Notice that the parameter τ_l, $l = 1,\dots,k$ is used for the minimum cluster size constraint. In other words, the cluster l is required to have at least τ_l points in it. This constraint is introduced to ensure that an adequate number of points falls into the each cluster to form multi-item groups. Therefore, undesired empty clusters, single-item clusters, and clusters with very few points in conventional k-means clustering are avoided at the end of the clustering process with this constraint.

The methodology used for constrained clustering can be summarized in the following steps (Demiriz et al. 2010a). Notice that the transactional dataset d is used for the association mining step and the constrained clustering is run on a different set of data, $\mathscr{X}$, in which the item properties can be represented. Constrained clustering is an iterative procedure, and s represents the iteration number in the algorithm below.

1. **Determine Must-Link Constraints.** Apriori algorithm is run on the transactional dataset d to generate frequent item pairs (2 itemsets) with a minimum support value of min_sup, and to determine the set of must-link constraints, i.e., $\{c_=(i, j)\}\, i, j = 1, \ldots, m$.
2. **Determine Cannot-Link Constraints.** The indirect association mining algorithm is run on d to generate itemsets that have pairwise *negative* association and to determine the set of cannot-link constraints, i.e., $\{c_{\neq}(i, j)\}\, i, j = 1, \ldots, m$.
3. **Run the Constrained Clustering.** Iterate the following steps (A, B, and C) to generate item clusters until the solution converges:

   A. **Cluster Assignment.** Let $Y_{i,l}^s$ be a solution to Lagrangian relaxation of Eq. 3 with $\mathscr{C}_{l,s}$ fixed.
   B. **Cluster Update.** Update $\mathscr{C}_{l,s+1}$ as follows:

$$\mathscr{C}_{l,s+1} = \begin{cases} \dfrac{\sum_{i=1}^m Y_{i,l}^s x_i}{\sum_{i=1}^m Y_{i,l}^s} & \text{if } \sum_{i=1}^m Y_{i,l}^s > 0, \\ \mathscr{C}_{l,s} & \text{otherwise.} \end{cases} \tag{5}$$

   C. $s \leftarrow s + 1$.

3 Multiple Forecasts in Retail

The basic idea in forecasting is to use the historical data to predict the future, assuming that the forecasting tool can model the underlying time series sufficiently and the structure of the time series does not change over the prediction period. However, a retailer may face more challenges due to the large number of items and store combinations, dynamic assortments, increased marketing activities, and the nature of the demand itself (Thomassey and Fiordaliso 2006).

Traditionally, forecasting the demand of the following types of items is harder than usual: slow-moving items, highly seasonal items, items with very short life spans (e.g., a few weeks), perishable items, items with the long lead-times, and items where marketing activities are present more than as usual (more frequent promotions). The common problematic characteristic of all these item types is the high variability of the demand being forecasted, i.e., demand volatility. One way of coping with the problem of high variability involves grouping the items either under the umbrella of traditional item categories (Shocker et al. 2004) or under some clustering scheme that might work better (Bunn and Vassilopoulos 1999; Kumar and Patel 2010). Grouping items can open ways to aggregate the sales of the items that fall into the corresponding groups. The basic idea of aggregation is to construct so-called item families, consisting of individual items with similar sales patterns (Bunn and Vassilopoulos 1999; Dekker et al. 2004). Seasonal indices may be found at the aggregate level, and these indices can be used later to forecast at the individual item level (Bunn and Vassilopoulos 1999; Dekker et al. 2004). However, the reason behind the grouping of the similar items in this work is not just for aggregation to reduce the variability but for regressing the item demands within each group by using the data from all the related items. This is indeed one of the novelties of our approach.

A survey of methods that combine multiple forecasts for a single item can be found in de Menezes et al. (2000). Combining multiple forecasts via the least square method was introduced in Clive (1984). The study compares the case where weighted combinations of forecasts when the sum of weights equals to 1, to that without any constraints on weights. It is shown that weighted combinations without any constraints on weights (i.e., coefficients of regression models) performed better on an econometric series (Clive 1984).

The idea behind combining multiple forecasts for an item is to reduce the variability with additional information coming from multiple forecasts without any sacrifice in bias. However, this is not true when forecasts belonging to the different items are combined (Kumar and Patel 2010). The aims of Kumar and Patel (2010) are to group the items that have similar forecasts and then to combine their forecasts to reduce the variability while not sacrificing the additional bias introduced by the combination. A heuristic method is proposed in Kumar and Patel (2010) to group similar items in terms of their next period forecasts.

Demand forecasting at stock keeping unit (SKU) level under price promotions is studied in Gür-Ali et al. (2009). Item hierarchy is used in Gür-Ali et al. (2009) to determine the related items that are incorporated into more complex models. Simple forecasting models work reasonably well under the conditions without any promotion effect. However, complex models with additional data preparation steps, such as including the information from the related items, performed better in the presence of promotions (Gür-Ali et al. 2009).

Item grouping should be limited to neither the traditional category definition nor the clusters found by similarity measures based on the sales figures. Complementary and substitute items can also form alternative groupings to improve forecasting accuracy and to reduce the variability. Gürhan Kök and Fisher (2007) use substitution effects to estimate demand functions better within an assortment. Therefore, approach in Gürhan Kök and Fisher (2007) enables determining the right products and their inventory levels for the purpose of assortment planning. The multinomial logit (MNL) model is the backbone of determining substitution probabilities in Gürhan Kök and Fisher (2007). In this work, we also assume that the price of a product is the most important determinant of its utility. Therefore, product prices are the crucial ingredient of the demand forecasting. Forecasting within the related item groups is both a form of variable selection method and a way of local learning, i.e., building local forecasting models to predict product demands better.

More recently, Johnson et al. (2014) introduce regression tree-based bagging models to determine the demand functions of so-called flash sales for an online fashion retailer. In this particular problem setting, scarcely available designers' apparel and accessories at the style level of the product hierarchy are sold in very limited time windows, i.e., events. Therefore, each event may contain limited inventory items. The lost sales are computed before applying the inventory-based demand constraints in Johnson et al. (2014) for a more accurate demand estimation. Finally, a multi-item price optimization model is then used for finding the optimal pricing decisions for all the competing styles. Since events last for a short period times, the price optimization is not designed for dynamic pricing, rather it is considered as one-time optimization. Moreover, Fisher et al. (2015) solve an online retailer's dynamic pricing problem to incorporate the competitive responses under demand constraints. Consumer choice models are applied by using competitors' stock-out information too.

4 Deterministic Dynamic Pricing Model

Markdown optimization is in most cases considered only for single items due to the complexity of underlying network (multiproduct) models—see Chap. 5 of Talluri and Van Ryzin (2004). On the other hand, existence of strong cross-elasticities among products and availability of data mining tools enable retail analysts to discover complementary effects which exist among thousands of items easily. Therefore, a new perspective can be easily introduced to retail domain to present the value of multi-item markdown optimization that incorporates product associations. In many cases, pricing decisions of an item are made by considering the prices within its category and competitors' prices in addition to demand and cost data of that particular item (Sung and Lee 2000). When it comes to conducting multi-item markdown optimization, one of the first questions that may arise in the minds of the practitioners is how to choose multiple products to determine a group of products. Intra-category product relations are sometimes easy to find and understand by the analysts in an ad hoc manner, but it might be very challenging to detect inter-category product relationships as search space might become very large to analyze by analysts. Therefore, usage of data mining techniques is needed to determine the multi-item groups. A simple clustering algorithm would be sufficient in most cases. However, product associations are also utilized in our framework to find the related items.

After multi-item groups are formed according to some clustering schemes, another issue arises to be resolved by practitioners: How to determine the underlying demand functions? We can construct linear demand functions between products within a multi-item group for this purpose. To give an example by following Chap. 2 of Shy (2008), let us assume there are two items: a and b. Demand functions of both items depend on the prices of the items in corresponding demand functions separately as follows:

$$D_a(p_a, p_b) = \beta_0^a + \beta_1^a p_a + \beta_2^a p_b$$
$$D_b(p_a, p_b) = \beta_0^b + \beta_1^b p_a + \beta_2^b p_b$$

Notice that $\beta_1^a < 0$ and $\beta_2^b < 0$ must hold in order for each item's demand to be inversely proportional to its price. Depending on the signs of β_2^a and β_1^b, product b and a may become complement or substitute in each other's demand function. Notice that they may have to satisfy the sign constraints accordingly in order to become proper complement or substitutes (Shy 2008). However, we did not enforce the sign constraints in our empirical study not to introduce complexities to the linear regression models. In addition to prices, time (i.e., week) has also been used as a variable in the demand function (see Eq. 6). With the inclusion of time, a problem may arise in terms of product life spans. Certainly, life spans of the products in a multi-item group should be similar and comparable. By definition, they have to overlap significantly to generate the item associations. Otherwise, we cannot speak of proper item associations. Therefore, overall item sales will have similar seasonal behaviors. To sanitize and limit our study, certain assumptions are made in Sect. 5 to pick proper items used in our experiments.

A simple deterministic multi-item markdown optimization model can be constructed by maximizing the total profit of a retailer subject to some capacity constraints (Talluri and Van Ryzin 2004). In our case, there exist limited product inventories without replenishment (as an assumption—it is often permissible to replenish a product during the retail season). Therefore, one of the decision variables would be initial inventory level besides our primary decision variable, price (p). Total profit may be a function of total revenue, salvage values, and inventory holding costs. We first introduce the parameters and the variables used in our model, and then the mathematical programming model is provided in Appendix. A similar model can also be found in Cosgun et al. (2013), but the deterministic model was never truly utilized as the multi-item relations were not properly represented in Cosgun et al. (2013).

Equation 6 represents the demand function as aforementioned. In addition to the fixed term (i.e., intercept, β_0^i) and product prices within the product group, the weekly demand depends on t, the time in weeks. Notice that regression models determine β coefficients within a product group. Assume there are n_l items within a group, then the regression models yield n_l sets of β coefficients. Consequently, Eqs. 6–11 are constraints related to the inventory levels. For example, Eq. 7 states that the end-of-week inventory of a particular product is equal to the beginning inventory level minus weekly sales of that product. Similarly, starting inventory level of the next time period is equal to starting inventory level of current time period minus the sales in the current week (see Eq. 11). Equations 12–15 practically enforce the positive demand constraint and its boundaries. Recall that binary decision variable r_{it} enforces the demand, D_{it} to be positive or equal to zero (see Eq. 12). Similarly, Constraint (13) enforces the upper limit of the demand to be equal to $wF D_{it}$ found by Eq. 6. Constraint (14) requires that the total sales of a particular product at a particular week cannot exceed the total demand of that product in that week.

Equations 16–22 specify the required boundaries of the product prices. For example, the price in the next period, p_{it+1}, should be smaller or equal to the current period price, p_{it} (see Eq. 16). Constraint (17) limits the price of a product at the very last week of the season to be greater than or equal to the salvage value of that product. Notice that the binary variable B_{it} indicates the markdown decision for the product i in period t. Therefore, a price difference as low as ε should exist between the consecutive periods (see Eq. 19). Equations 23 and 24 determine the number markdowns throughout the season. Inventory holding costs are calculated by Eqs. 25 and 26. At the very end of mathematical program, the boundary conditions of the decision variables are given.

5 Empirical Study

In this section, the applicability of our proposed framework is demonstrated through a real-world dataset, used in our earlier work (Demiriz et al. 2010b, 2011) as well, involving the store-level retail sales data originating from a leading apparel retail chain in Turkey. In our work, there was access to the complete sales, inventory, and transshipment data belonging to a single merchandise group (men's clothes line) for the 2007 summer season, coming from all the stores of the retail chain. Some summary statistics from the dataset are given in Table 2.

The detailed explanation of the dataset which originates from retail domain can be found in our earlier work in Demiriz et al. (2010b, 2011). Hierarchical structure of product tree was explained in Demiriz et al. (2010b, 2011) in detail. We simply state here that our analysis has been done at model level instead of SKU level.

Table 2 A high-level summary of retail data

Number of	
SKUs	8,807
Model items	716
Transactions	2,753,260
Items sold	4,376,886
Stores	172

5.1 Finding-Related Item Groups

Since our aim is to construct the linear regression models to forecast the weekly demands of the items by means of time (i.e., week), the prices of items at the beginning of the each week, there is no need to use the complete dataset, especially the data for the items whose prices seldom fluctuate. This is especially true for the basic items which are available throughout the season at relatively lower prices without major price markdowns. Therefore, a filtering scheme is used for determining a subset of items where item prices may have greater effects. Items are filtered further according to their initial price, the difference between the initial prices and the end of the season prices, number of weeks sold (i.e., item lifetime), and the coefficient of variation of the price to select a suitable subset. The filtering scheme has yielded a subset that consists of 55 items. These 55 items have 41 positive and 38 negative pairwise associations in between them.

Technically, a hierarchical constrained clustering method (Gilpin and Davidson 2011) may be used to cluster items that have an inherent product hierarchy. However, a mathematical programming approach is preferred in this work over intelligent search techniques that may require solving a more complicated constraint satisfaction problem. Therefore, three methods described in Sect. 2 are used to group those 55 items. The results are reported in Table 3. Item IDs that fall into the corresponding groups are given in Table 3. Note that the order of the item IDs is equivalent to the lexicographic order of the item codes (names).

Table 3 Item groups and their members found by different methods

Grp	Associated groups	k-means clustering	Constrained clustering
#1	10, 39, 43, 47	43	4, 7, 17, 18, 25, 27
#2	8, 22, 34, 35	10, 22, 39	24, 33, 49
#3	1, 49, 52, 54, 55	41	1, 15, 54, 55
#4	46, 48	46	13, 14, 16, 26, 29, 36, 51
#5	23, 41	2, 9, 11, 26, 31, 32, 37, 47, 50, 52, 53	19, 28, 50, 52
#6	19, 28, 32, 51	3, 4, 5, 6, 7, 12, 17 ,18, 25, 27, 38, 40, 42, 44, 45	3, 5, 6 ,12, 42, 44, 45
#7	4, 17, 18	1, 23, 48, 54	8, 21, 34, 35
#8		13, 30, 34, 35	2, 10, 22, 38, 39, 40, 43, 46, 48
#9		24, 33, 49	23, 30, 41
#10		8, 14, 15, 16, 19, 20, 21, 28, 29, 36, 51, 55	9, 11, 20, 31, 32, 37, 47, 53

Table 4 Comparison of k-means and constrained clustering

Comparison	k-means clustering	Constrained clustering
Objective value	39,566	8,049,714
Min. cluster size violations	3	None
Must-link violations	24	7
Cannot-link violations	11	1

As expected, the heuristic grouping scheme is not able to group all the items; i.e., some of the items are left alone by this heuristic. Only 24 out of 55 items are grouped by this heuristic into seven different groups. The weekly demand of the remaining 31 items can be modeled based on the individual item's price and time (i.e., value of the week). In other words, 31 items can be considered as singleton clusters. Notice that there are two groups that are composed of two items. Basically, two items form a multi-item group, but the relations are restricted to the pairwise relations. If the pairwise relations are sufficient, we could simply use the pairwise item associations without any further grouping. However, the strength of multi-item can be seen at larger cluster sizes due to higher number of available predictors within larger clusters.

On the other hand, k-means clustering ends up with three clusters that have only one item in each cluster (i.e., singleton cluster). This is not a desirable outcome. The number of clusters is set to be ten for both k-means and constrained clustering as seen in Table 3. The cluster size changes between 1 and 15 for the k-means clustering. The minimum cluster size, the parameter τ, is set to be three in constrained clustering. Therefore, the cluster size changes between 3 and 12 for the constrained clustering. Both k-means and the constrained clustering are implemented by using IBM ILOG OPL Studio.[1] Interested readers can consult with accompanying models provided in Demiriz et al. (2010a).

The results from k-means and the constrained clustering methods are compared in Table 4. Note that the penalty parameters λ^+ and λ^- of the Lagrangian relaxation are both set to 1,000,000 in the constrained clustering experiments. Therefore, there is a significant jump in the objective value of the constrained clustering in which eight must-link and cannot-link violations occurred totally. On the other hand, the total number of violations is 35 for the k-means clustering. The actual objective value of the k-means clustering should be around $35M$ when the constraints are considered.

---

[1] https://developer.ibm.com/academic/.

5.2 Conducting Multivariate Regression Analysis Within Item Groups

As reported in Gür-Ali et al. (2009), simple prediction models may yield competitive results in comparison to the complex ones. Therefore, relatively simple regression models are implemented in this subsection to forecast the item demands. A multivariate linear regression (Ordinary Least Squares—OLS) model can simply be posed as

$$y = X\beta + \varepsilon$$

where $y \in \mathbb{R}^n$ is the response (dependent) variable, the matrix $X \in \mathbb{R}^{n \times d}$ represents the independent variables, $\beta \in \mathbb{R}^d$ are the regression parameters (coefficients), and the error terms $\varepsilon = (\varepsilon_1, \ldots, \varepsilon_n)$ are i.i.d. with mean 0 and variance σ^2.

In a retailing context, the weekly prices and the time (i.e., the number of weeks passed since the item was first sold—shelf time) can easily be used for modeling weekly item demands via regression models. Since there are m items considered, m different models will be built and the total number of variables is $2m + 1$ in the reduced dataset: one set of m variables for the total demand of the items (i.e., response variables) and the price of the items at the beginning of the week in addition to the time variable (i.e., corresponding week). Note that the reduced dataset might have observations at most as many as the number of weeks in a season plus a few more weeks for the clearance period. In other words, the regression models might have 30 observations at most in a season. If there are a lot of items that are considered in the modeling, then the variable selection becomes more critical. For example, let us assume the best four variables are sought to model the weekly demand. Therefore, search space needs to be covered in at most $\binom{m+1}{4}$ steps.

The traditional variable selection methods such as forward selection, backward elimination, and stepwise regression are common in practice. The results from such selection methods might be arbitrary and sub-optimal in some cases. One of the variable selection methods that is "the best subset selection" method is proven to be NP-hard in terms of computational complexity (Xie and Zeng 2010) and it returns the optimal subset of the variables. However, it is not practical due to its complexity, especially for the high-dimensional datasets.

As mentioned in Sect. 2, the models based on the related item groups achieve an indirect form of variable selection method since the models are built only within the related item groups by regressing one item's demand against the remaining items' price data as well as the focal item's price and the time (i.e., week). Notice that the time variable represents the shelf **time** of the focal product. In other words, the time passed since the item was first sold in any store. Planning horizons may slightly differ within the multi-item group. All variable selection methods will have less complexity when they are applied directly within the related item groups. For example, assuming that m items are evenly distributed among the related item groups that satisfy the minimum cluster size constraint then the search space will cover $k\binom{\frac{m}{k}+1}{4}$ steps for the best four subset variable selection scheme for all the clusters.

Once the related item groups are found as in the previous subsection, regression models can be built. Four different sets of experiments are conducted, and the results are summarized in Table 5. Besides the regression results from three different item groupings, the linear regression models are built based on the single items to compare our framework with existing approaches. Due to the limited number of observations for each item, it is not possible to utilize hold-out samples to verify regression models. Therefore, Table 5 reports traditional statistics like average R^2 (coefficient of determination) and average root mean squared error (RMSE) across all the methods. R^2 can be computed as:

$$R^2 = 1 - \frac{SS_{res}}{SS_{tot}}$$

where the residual sum of squares (SS_{res}) is calculated by $\sum_i (y_i - \hat{y}_i)^2$ and the total sum of squares (SS_{tot}) is calculated by $\sum_i (y_i - \bar{y}_i)^2$. Note that y is the dependent variable in the regression model, and $\hat{y}$ and $\bar{y}$ are the predicted and the mean values of the dependent variable, respectively. RMSE is technically equal to $\sqrt{SS_{res}/n}$.

For computing the average values of associated group heuristic, we used single-item model results if that particular point was not assigned to any group. It should be noted that we can utilize a variable selection method such as stepwise regression within each group. However, such approach might yield sparse regression models. The full regression models were preferred instead. In other words, each demand model of the items within a group depends on all the item prices within that particular group.

As expected, the single-item models cannot fit the data as good as the other methods due to the limited explanatory capacity. Recall that single-item model uses only the corresponding item's price and the time (i.e., week). The idea of grouping items definitely helps the forecasting step as the results indicate higher R^2s and lower RMSEs for the multi-item groups. The best grouping scheme turned out to be k-means clustering for regression results in our experiments. In other words, k-means clustering yielded the best average R^2 and RMSE. We need to consider that the constrained clustering may result in evenly distributed clusters—in terms of number of points falling into each cluster. In many cases, k-means clustering may result in lopsided clusters meaning that some clusters may have too many points in them but the rest of the clusters may have very few points in them. This may result in very poor forecasting, variable selection, and overfitting problems in large clusters. It may also result in sub-optimal markdown price optimization. We used only 55 products in our experiments, but there might be thousands of items in practice and this could create the overfitting problem easily for the large clusters.

5.3 Implementing Deterministic Multi-item Markdown Optimization Model

The mathematical programming model introduced in Sect. 4 was solved for multi-item groups formed by three grouping schemes and for single-item case in our empirical study. The results are reported in the last column of Table 5. GAMS modeling language was used in this part of the study to express the mathematical programming model. DICOPT solver was used to run the models. A SAS macro was written to generate GAMS model files automatically based on the results of the regression models mentioned in the previous subsection. Some model parameters and boundaries of decision variables are determined based on the real data. We set the salvage value (sv_i) of all the products to be three and the unit cost of a product in inventory for a week (h_{it}) to be 0.001 for all cases. Except very few cases, all models were solved to optimality. Otherwise, sub-optimal results were returned at the run-time limit by using GAMS option of 300 s.

One of the issues regarding the implementation was to determine shelf lives for all the items within a group. Although items might have different shelf lives, we picked the longest shelf life in the group as the common shelf life for all the items within a group. Since we regress item demands within a multi-item group with each other, we need to use a common planning horizon for all of them. Notice that the single-item model requires the usage of real shelf life of the corresponding product. In a way, shelf life determines how many observations used in the regression models.

As seen from Table 5 that constrained clustering resulted in the best profit for the firm. Nevertheless, multi-item markdown optimization model has generated higher profit margins compared to single-item model. Notice that single-item model can be considered as a special case of multi-item model where n_l is equal to 1. We also reported in Table 6 a detailed profit comparison between k-means and constrained clustering methods at cluster level. Cluster compositions are previously reported in Table 3. Notice that optimization model for the Cluster 8 of constrained clustering resulted a sub-optimal solution. It might have yielded a bit more profit than the optimal solution. However, even it is a bit lower, the total profit of multi-item markdown optimization of constrained clustering is still better or comparable with k-means clustering. Sub-optimal solutions could be yielded for k-means clusters as well.

Table 5 Experimental results

Underlying method	Average R^2	Average RMSE	Total profit
Single-item models	0.65	169.8	15,139,313
Associated group heuristic	0.70	159.4	17,137,381
k-means clustering	0.88	118.2	21,181,775
Constrained clustering	0.83	131.3	21,809,985

Table 6 Detailed profit comparison between k-means and constrained clustering

Cluster	1	2	3	4	5	6	7	8	9	10
k-means	470,838	1,751,127	245,510	341,465	3,888,529	4,968,769	1,688,578	1,971,883	930,794	4,924,282
Constrained	2,290,819	930,794	2,545,137	2,545,137	2,605,991	1,082,704	1,854,339	5,202,583	594,787	2,157,695

6 Concluding Remarks

A retail forecasting framework is introduced in this chapter, based on the item groups derived from the similarities of the sales figures and the item association constraints extracted from the complementary and the substitution effects. Constrained clustering is used for ensuring that the item association constraints are satisfied as much as possible, while maximizing the similarities based on the sales figures within each group. Item demands are estimated by using linear regression models within each group, where the prices and the time (i.e., week) are used as the explanatory variables. Building regression models within each item group (cluster) can be considered as a form of variable selection method and local learning.

Cannibalization is an important issue for markdown optimization. We utilized both must-link and cannot-link constraints to form multi-item groups to minimize adverse effects of cannibalization. An alternative way could be using only must-link constraints for finding multi-item groups. Linear regression models are used throughout the chapter for their simplicity to compare various grouping schemes shown in this chapter. Our results indicate that grouping the items and then forecasting the demand within each group is superior to the single-item demand forecasting. More elaborate and more advanced techniques such as neural networks, SVM regression and even autoregressive models could be utilized in an extended study. Similar results should be expected across the modeling alternatives since grouping plays an important role as a variable selection method. Our primary goal was to show the applicability of the idea that grouping items before conducting a forecasting study improves the forecasting results within these groups, even though aggregating the item demands was not considered in this study. We also showed that grouping products first and then applying multi-item markdown optimization improves the profit margin of the firms compared to separate markdown optimization for each product. This was achieved through an empirical study with a real-world dataset from an apparel retail chain.

Appendix

Multi-item Markdown Optimization Model

The parameters of the model are given below:

n	number of products
n_l	number of products in cluster l
T	number of weeks in the season
T_l	planning horizon (weeks) for cluster l
ε	very small price value (such as 1 cent), which will allow to differentiate between two prices
IS_i	initial inventory for product i (at the beginning of the season)

IP_i	initial price of product i (at the beginning of the season)
$\beta_0^i, \ldots, \beta_{n_l}^i$	coefficients for the products in cluster l that represent how they contribute to the sales of product i
h_{it}	unit cost of holding product i in inventory during week t
sV_i	salvage value/price of product i (can be fixed as parameters, if desired)
lMS	maximum number of markdowns throughout a season (optional, fixed by the retail managers)
M	a large number

The variables of the model are given below:

p_{it}	price of product i during week t
B_{it}	binary variable that indicates whether a markdown is applied to product i in week t (1 if markdown was applied)
r_{it}	binary variable that indicates whether the demand forecast for product i in week t is positive (1 if demand forecast is positive)
wFD_{it}	demand forecast for product i for week t (even though this can be a general function of p_{it} and time, constraint (1) models the special case where it is a linear function of p_{it})
D_{it}	*positive* demand forecast for product i for week t
S_{it}	sales of product i in week t
fS_i	number of units of product i left in inventory at the end of the season
TS_i	total sales of product i throughout the season
wIS_{it}	initial inventory of product i in week t
wFS_{it}	ending inventory of product i in week t
hC_{it}	total cost of inventory for product i during week t
THC_i	total cost of inventory for product i throughout the season
nMS_i	number of markdowns applied for product i

$$\max \sum_{i=1}^{n_l} \sum_{t=1}^{T_l} p_{it} S_{it} + \sum_{i=1}^{n_l} s V_i fS_i - \sum_{i=1}^{n_l} \sum_{t=1}^{T_l} h_{it} wFS_{it}$$

$$s.t.$$

$$wFD_{it} = \beta_0^i + \sum_{j=1}^{n_l} \beta_j^i p_{jt} + \beta_{n_l+1}^i t \quad \forall i \tag{6}$$

$$wFS_{it} = wIS_{it} - S_{it} \quad \forall i, \forall t \tag{7}$$

$$wIS_{i1} = IS_i \quad \forall i \tag{8}$$

$$TS_i = \sum_{t=1}^{T_l} S_{it} \quad \forall i \tag{9}$$

$$fS_i = IS_i - TS_i \quad \forall i \tag{10}$$

$$wIS_{it+1} = wIS_{it} - S_{it} \quad \forall i, \forall t \tag{11}$$

$$D_{it} \leq Mr_{it} \quad \forall i, \forall t \tag{12}$$

$$D_{it} - wFD_{it} \leq M(1 - r_{it}) \quad \forall i, \forall t \tag{13}$$

$$S_{it} \leq D_{it} \quad \forall i, \forall t \tag{14}$$

$$wIS_{it} \geq S_{it} \quad \forall i, \forall t \tag{15}$$

$$p_{it+1} \leq p_{it} \quad \forall i, \forall t \tag{16}$$

$$p_{iT} \geq sV_i \quad \forall i \tag{17}$$

$$p_{i1} \leq IP_i \quad \forall i \tag{18}$$

$$p_{it+1} - p_{it} + \varepsilon \leq M(1 - B_{it+1}) \quad \forall i, \forall t \tag{19}$$

$$p_{i1} - IP_i + \varepsilon \leq M(1 - B_{i1}) \quad \forall i \tag{20}$$

$$p_{it+1} - p_{it} + MB_{it+1} \geq 0 \quad \forall i, \forall t \tag{21}$$

$$p_{i1} - IP_i + MB_{i1} \geq 0 \quad \forall i \tag{22}$$

$$nMS_i = \sum_{t=1}^{T_i} B_{it} \quad \forall i \tag{23}$$

$$nMS_i \leq lMS \quad \forall i \tag{24}$$

$$hC_{it} = h_{it} wFS_{it} \quad \forall i, \forall t \tag{25}$$

$$THC_i = \sum_{t=1}^{T_i} hC_{it} \quad \forall i \tag{26}$$

$$p_{it} \geq 0, B_{it} \in \{0, 1\}, r_{it} \in \{0, 1\} \quad \forall i, \forall t$$

$$sV_i \geq 0, fS_i \geq 0, TS_i \geq 0, THC_i \geq 0 \quad \forall i$$

$$D_{it} \geq 0, hC_{it} \geq 0, wFD_{it} \text{ unbounded} \quad \forall i, \forall t$$

$$S_{it} \geq 0, wFS_{it} \geq 0, wIS_{it} \geq 0 \quad \forall i, \forall t$$

$$nMS_i \geq 0 \text{ and integer} \quad \forall i$$

References

Basu S, Davidson I, Wagstaff KL (eds) (2008) Constrained clustering: advances in algorithms, theory, and applications. Chapman and Hall/CRC

Bradley PS, Mangasarian OL, Street WN (1997) Clustering via concave minimization. In: Mozer MC, Jordan MI, Petsche T (eds) Advances in neural information processing systems. MIT Press, Cambridge, MA, pp 368–374. ftp://ftp.cs.wisc.edu/math-prog/tech-reports/96-03.ps.Z

Bunn DW, Vassilopoulos AI (1999) Comparison of seasonal estimation methods in multi-item short-term forecasting. Int J Forecast 15(4):431–443

Cosgun O, Kula U, Kahraman C (2013) Markdown optimization via approximate dynamic programming. Int J Comput Intell Syst 6(1):64–78

de Menezes LM, Bunn DW, Taylor JW (2000) Review of guidelines for the use of combined forecasts. Eur J Oper Res 120(1):190–204

Dekimpe MG, Hanssens DM (2000) Time-series models in marketing: past, present and future. Int J Res Mark 17:183–193

Dekker M, van Donselaar K, Ouwehand P (2004) How to use aggregation and combined forecasting to improve seasonal demand forecasts. Int J Prod Econ 90:151–167

Demiriz A (2015) Using pairwise associations for multi-item markdown optimisation. Int J Syst Sci Oper Logist 1–14. https://doi.org/10.1080/23302674.2015.1094835

Demiriz A, Cihan A, Kula U (2009) Analyzing price data to determine positive and negative product associations. In: Leung C, Lee M, Chan J (eds) Neural information processing. LNCS, vol 5863. Springer, pp 846–855

Demiriz A, Ertek G, Atan T, Kula U (2011) Re-mining item associations: methodology and a case study in apparel retailing. Decis Support Syst 52(1):284–293

Demiriz A, Ekizoglu B, Kula U (2010a) Clustering products under pairwise positive and negative association constraints in retailing. In: Proceedings of ANNIE 2010, St. Louis, MO, USA, November, 2010

Demiriz A, Ertek G, Atan T, Kula U (2010b) Re-mining positive and negative association mining results. In: Perner P (ed) Advances in data mining. applications and theoretical aspects, vol 6171 of LNCS. Springer, pp 101–114

Demiriz A, Bennett KP, Bradley PS (2018) Using assignment constraints to avoid empty clusters in k-means clustering. In: Basu et al (eds) [1], pp 203–222

Fisher M, Gallino S, Li J (2015) Competition-based dynamic pricing in online retailing: a methodology validated with field experiments. Technical report, Ross School of Business, University of Michigan, January 2015

Fu T (2011) A review on time series data mining. Eng Appl Artif Intell 24(1):164–181

Gardner ES Jr (2006) Exponential smoothing: the state of the art-Part II. Int J Forecast 22(4):637–666

Gilpin S, Davidson I (2011) Incorporating sat solvers into hierarchical clustering algorithms: an efficient and flexible approach. In: Proceedings of the 17th ACM SIGKDD international conference on knowledge discovery and data mining, KDD '11, New York, NY, USA, 2011. ACM, pp 1136–1144

Granger CWJ, Ramanathan R (1984) Improved methods of combining forecasts. J Forecast 3(2):197–204

Gür-Ali Ö, Sayın S, van Woensel T, Fransoo J (2009) SKU demand forecasting in the presence of promotions. Expert Syst Appl 36(10):12340–12348

Gürhan Kök A, Fisher ML (2007) Demand estimation and assortment optimization under substitution: methodology and application. Oper Res 55(6):1001–1021

Hasan M, Salem S, Zaki M (2011) SimClus: an effective algorithm for clustering with a lower bound on similarity. Knowl Inf Syst 28:665–685. https://doi.org/10.1007/s10115-010-0360-6

Johnson K, Lee BHA, Simchi-Levi D (2014) Analytics for an online retailer: demand forecasting and price optimization. Technical report, MIT

Kumar M, Patel N (2010) Using clustering to improve sales forecasts in retail merchandising. Ann Oper Res 174:33–46. https://doi.org/10.1007/s10479-008-0417-z

Lange T, Law MH, Jain AK, Buhmann JM (2018) Clustering with constraints: a mean-field approximation perspective. In: Basu et al (eds) [1], pp 91–122

Lu Z, Leen TK (2018) Pairwise constraints as priors in probabilistic clustering. In: Basu et al (eds) [1], pp 59–90

Nijs VR, Dekimpe MG, Steenkamp J-BEM, Hanssens DM (2001) The category-demand effects of price promotions. Mark Sci 20(1):1–22

Pensa RG, Robardet C, Boulicaut JF (2018) Constraint-driven co-clustering of 0/1 data. In Basu et al. [1], pp 123–148

Shocker AD, Bayus BL, Namwoon K (2004) Product complements and substitutes in the real world: the relevance of "other products". J Mark 68:28–40

Shy O (2008) How to price: a guide to pricing techniques and yield management. Cambridge University Press

Sung NH, Lee JK (2000) Knowledge assisted dynamic pricing for large-scale retailers. Decis Support Syst 28(4):347–363

Talluri KT, Van Ryzin G (2004) The theory and practice of revenue management. Springer, International Series in Operations Research & Management Science

Tan P-N, Kumar V, Kuno H (2001) Using SAS for mining indirect associations in data. In: Western Users of SAS software conference

Thomassey S, Fiordaliso A (2006) A hybrid sales forecasting system based on clustering and decision trees. Decis Support Syst 42(1):408–421

Vindevogel B, den Poel Van D, Wets G (2005) Why promotion strategies based on market basket analysis do not work. Expert Syst Appl 28(3):583–590

Wu X, Kumar V, Ross Quinlan J, Ghosh J, Yang Q, Motoda H, McLachlan G, Ng A, Liu B, Yu P, Zhou Z-H, Steinbach M, Hand D, Steinberg D (2008) Top 10 algorithms in data mining. Knowl Inf Syst 14:1–37 (2008). https://doi.org/10.1007/s10115-007-0114-2

Xie J, Zeng L (2010) Group variable selection methods and their applications in analysis of genomic data. In: Feng J, Fu W, Sun F (eds) Frontiers in computational and systems biology, vol 15. Computational biology. Springer, London, pp 231–248

Social Media Analytics for Decision Support in Fashion Buying Processes

Samaneh Beheshti-Kashi, Michael Lütjen and Klaus-Dieter Thoben

Abstract The Web 2.0 and the emergence of numerous social media services enable individual users to publish and share information on the one hand and to discuss diverse topics online on the other hand. Accordingly, different research streams have emerged in order to tackle the diverse phenomena related to social media. Social media analytics as an interdisciplinary research field has arisen and integrates the different approaches of structural attributes, opinion/sentiment-related as well as topic/trend-related approaches. This research follows topic- and trend-related approaches with the methods content and trend analysis on social media text data. These methods might be applied on different domains including the fashion industry. This research focusses on the fashion industry for three reasons. Firstly, this industry is a highly consumer-oriented industry, and these consumers themselves are the users of social media services. Secondly, the industry faces challenges in meeting the demand of the customer on time. Thirdly, in the last years, fashion blogs have gained increased relevance from the consumers and the industry. Accordingly, the fashion blogs may contain information for supporting decision maker in the industry, to perform their tasks such as meeting the demand with a lower degree of uncertainty. The objective of this chapter is to explore the potential added value of social media analytics for fashion buying processes, not only by presenting an abstract approach,

S. Beheshti-Kashi (✉)
International Graduate School for Dynamics in Logistics (IGS), University of Bremen, Hochschulring 20, 28359 Bremen, Germany
e-mail: bek@biba.uni-bremen.de

M. Lütjen
BIBA - Bremer Institut für Produktion und Logistik GmbH, University of Bremen, Hochschulring 20, 28359 Bremen, Germany
e-mail: ltj@biba.uni-bremen.de

K.-D. Thoben
Faculty of Production Engineering, University of Bremen, Badgasteiner Straße 1, 28359 Bremen, Germany
e-mail: tho@biba.uni-bremen.de

© Springer Nature Singapore Pte Ltd. 2018
S. Thomassey and X. Zeng (eds.), *Artificial Intelligence for Fashion Industry in the Big Data Era*, Springer Series in Fashion Business,
https://doi.org/10.1007/978-981-13-0080-6_5

but more by conducting experimental analyses on a fashion blog corpus covering a 5 year time period. Based on the topic detection and tracking research which origins from the intelligent information retrieval, a research approach is presented by integrating a text mining process, on the detecting and tracking of fashion features and topics in the blog corpus. A fashion topic may refer to different features such as a colour, silhouette or style. While for the topic detection task, the feature colour is focussed, the topic tracking includes topics on silhouette, style, colour and decorative applications. The analyses have shown that it is possible to detect single colour and co-occurred colour occurrences. In addition, it was demonstrated that it is possible to track fashion topics over a 5 year time period in a fashion post corpus. The fashion buyer might have an added value for his activities by quantifying the individual perceptions through the application of the presented approach.

Keywords Social media analytics · Text mining · Fashion buying · Fashion blogs · Topic detection · Topic tracking

1 Introduction

Within the last years, the Web 2.0 has enabled the emergence of numerous social media applications such as social network sites, blogs or microblogging services (Kaplan and Haenlein 2010). Every ordinary user with access to the Internet has the possibility to enter different social media applications. These users produce a large amount of unstructured data in different formats. In order to meet the demand of their customers, companies require capturing their preferences. In former times, they have been dependent on customer surveys or focus groups. Nowadays, they have direct access to customer opinions through those social media applications. This mainly freely accessible information has different formats. Most of these posts are unstructured or semi-structured such as free text, images and videos, which require adequate pre-processing and storage before applying advanced analytics. This chapter focusses on the exploitation of textual data via text mining which has gained an increased relevance within the last years due to the increase of published online data. Text mining has evolved as an interdisciplinary research field, applying methods mainly from the neighbouring fields such as machine learning, statistics, computational linguistics, library and information sciences, databases and data mining, where they apply concepts from the field of artificial intelligence (Miner et al. 2012).

The usage of social media text data might be highly relevant for various industries. However, this research considers the fashion industry as a suitable field of application due to several factors. Firstly, this industry is a highly consumer-oriented industry, and these consumers are themselves the users of social media services. Secondly, the industry faces challenges in meeting the demand of the customer on time. Thirdly, in the last years, fashion blogs have gained increased relevance from the consumers and the industry. Along a fashion supply chain, different stakeholder and processes might have interest in the usage of those blog data. However, a fashion buyer takes a crucial role in the success of the whole company (Wong 2013). He maintains collaborations with different other roles. In particular, he has to work deeply with designers, suppliers and merchandisers (Shaw and Koumbis 2013). The fashion buyer has different tasks, activities and responsibilities. To him basic task belongs amongst others to analyse the behaviour of the target groups. He undertakes trend monitoring activities such as formal and informal market research in order to react quickly to market changes (Jackson and Shaw 2001) which are often caused through the difficulty of predicting the customers' needs. The final objective is to select a collection for a particular time period. Prior to the market launching, the buyer has also to identify the appropriate prices and the volume which has to be produced. The goal is to sell as much as possible from the collection at the set price during that particular time period. If it is not possible to sell a huge number of a particular article, it often has to go on reduction which often results in losses for the company. For the companies, it is crucial to avoid the stages of stock-overs and also stock-out inventories. Therefore, the role of the buyer and his impact is highly crucial for the whole company and for the whole supply chain. For this reason, the current research is focussing on the buyer/buying team and his/their decision-making processes.

The objective of this chapter is to illustrate an approach within the social media analytics framework (Fig. 1), on how to deal with fashion blog posts by following a text mining procedure model, making experimental analyses of them, in order to finally discuss their actual added value by applying on the fashion buying scenario. The content of the results of the experimental analysis is not in the focus since the goal is not to identify certain content but to exemplarily illustrate the potential of analysing the blog posts. Therefore, the discussion part is focussed on the buyer's benefits.

The rest of this chapter is organized as follows. Section 2 gives an overview of social media and text mining. This is followed by the illustration of the research approach, displayed in Sect. 3. The results of the experimental analyses are presented in Sect. 4. Section 5 is dedicated to the discussion on the buyers' benefits from the analysis of the fashion posts. The chapter will conclude with a conclusion and outlook on the future research steps.

2 Theoretical Background

As aforementioned, the relevance of unstructured data has increased within the last years. Therefore, the necessity for automatic text processing and text mining approaches has accordingly increased. The following sections will mainly focus on social media and text mining processes.

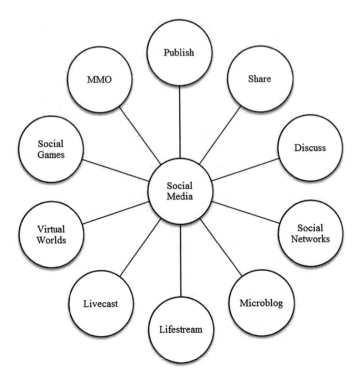

Fig. 1 Social media landscape (Power and Phillips-Wren 2012)

2.1   Social Media

With the rise of the Web 2.0 and the emerging technologies, the ordinary user gained a new role: he is an active and producing entity and not anymore passive and purely consuming. For this role, the literature introduced the term producer (Bruns 2006). Especially, fashion is a widely discussed topic in the communities and many fashion blogs publish different fashion-related topics. Kaplan and Haenlein (2010) define social media as a group of Internet-based applications that build on the ideological and technological foundations of Web 2.0 and that allow the creation and exchange of user-generated content. A further approach for defining social media is presented by Power and Phillips-Wren (2012) with the social media landscape illustrated in Fig. 1. Furthermore, in this discussion often the term of social software is mentioned in the literature (Alby 2007).

Schmidt (2009) makes a detailed classification and suggests the division of social software in (1) personal publishing, (2) platforms, (3) information management tools and wikis. Within the category of personal publishing, weblogs and microblogging services are located. Platforms are divided into multimedia platforms such as YouTube and social networking platforms such as Facebook. Social bookmarking services such as delicious are an example of information management tools. The wiki technology is used, for instance, for Wikipedia. Since this paper focusses on weblogs and forums, in the following both formats are described more in detail.

Personal publishing includes weblogs and microblogging services. A weblog is "a website that contains an online personal journal with reflections, comments, and often hyperlinks provided by the writer;" (Merriam Webster). In general, blogs can be differentiated in different types: private blogs, corporate blogs or media blogs (Zerfaß and Bogosyan 2007). Though, within the last years, a large number of lifestyle and fashion blogs have emerged.

The fashion blogs are the biggest group of the blogosphere (Halvorsen et al. 2013). Halvorsen et al. (2013) consider fashion blogs as blogs that focus on fashion brands, trends, products, e-commerce, street style and personal style. Fashion blogger has gained increased relevance from the fashion industry and is increasingly considered as influencers. The industry has realized the actual impact from the blogger on their reader which might be a potential customer of fashion companies. Accordingly, the companies have developed different strategies for cooperating with fashion bloggers in order to be able to use their reach out to the readers of the bloggers. They may invite them to shop openings or arrange specified events for the bloggers. They may send the products and hope to be written about the products on the blogs (Kulmala et al. 2014). These posts are often marked as sponsored posts from the blogger. Moreover, the blogger might include affiliate links on their blogs. The actual influence of the blogger might be measured through different tools. Consequently, companies are able to monitor their cooperation with the bloggers. The fashion blogs themselves consist of different categories such as fashion, outfits or travelling. Each post usually has a title, date, pictures and several textual paragraphs. For the purpose of this research, the posts which have been collected are mainly from fashion-related categories such as fashion or outfits. A first approach to the usage of fashion blogs in fashion processes was suggested by Rickmann and Cosenza (2007). They proposed a weblog text trending approach and focus more catching the actual "buzz" from the fashion posts. They conclude that for the actual usage of fashion blogs for trend forecasting purposes different changes/development are needed such as *rich accumulation of fashion communication* or the acceptance of fashion bloggers by the marketers. Within the last years, both developments have occurred. Further research on blogs and fashion processes is more focussed on their usage for influencing consumer behaviour such as in Halvorsen et al. (2013) or comparing brand associations from blogs and textual data from published by fashion companies themselves in Crawford Camiciottoli et al. (2014).

Textual data published on blogs and other social media have particular attributes such as time sensitivity, short length, unstructured phrases and abundant information. In particular, the unstructured format of the published content has been challenging for a long time, to tap into this valuable source, since it is not readable and understandable by machines (Hu and Liu 2012). However, due to the advances in text mining techniques a more efficient exploitation of the unstructured text data is possible. In order to gain some knowledge, the data has to pre-processed and transformed into structured data. This exploitation of high volume of data is possible through text mining methods, which is presented in the following chapter.

2.2 Text Mining

Text Mining and text analytics are broad umbrella terms describing a range of technologies for analysing and processing semi-structured and unstructured text data. The unifying theme behind each of these technologies is the need to "turn text into numbers" so powerful algorithms can be applied to large document databases. Converting text into a structured, numerical format and applying analytical algorithms require knowing how to both use and combine techniques for handling text, ranging from individual words to documents to entire document databases (Miner et al. 2012: 30). Newer trends in text mining such as social network analysis, multilingual text mining, spam classification, use of k-means clustering to group documents, anomaly detection, trend detection and analysis of streaming text data are described in Berry and Kogan (2010).

Miner et al. (2012) define seven practice areas of text mining: search and information retrieval (IR), document clustering, document classification, Web mining, information extraction (IE), natural language processing (NLP) and concept extraction. *The seven practice areas overlap considerably since many practical text mining tasks sit at the intersection of multiple practice areas.* However, for exploiting the textual data and generating an added value of this huge amount of data, several processes have to be conducted. The literature describes various process models or frameworks of text mining. Most text mining models suggested in the literature are either custom-designed for the presented case or highly generic illustrating merely the general functionality of text mining (Schieber and Hilbert 2014). In contrast, Schieber and Hilbert (2014) suggested a generic process model for text mining purposes. This process model was selected in this research since it is regarded as a detailed and exhaustive text mining model.

The model consists of six phases. The phases are named task definition, document selection and analysis, document preparation or document processing, application of text mining methods, evaluation of the results, derivation of actions and application of results.

Due to the unstructured nature of the textual documents, in particular, the document processing phase is a relevant phase to be conducted, in order to be able to apply classical data mining methods. Therefore, the document processing phase is focussed strongly in the following. The sub-processes of the document processing phase are depicted in Fig. 2.

The filtering, lexical, syntactic and semantic processing provides a repertoire of methods which are followed to achieve certain objectives. In the case of word filtering, for instance, so-called stop word lists which include articles or prepositions are used, in order to filter irrelevant words and reduce the number of terms in a given

Fig. 2 Document
processing steps

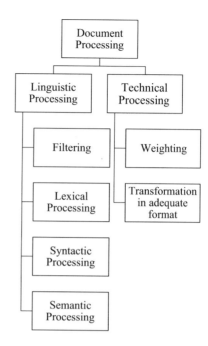

text corpus. Table 1 summarizes the objective of the sub-processes of the linguistic processing with the main tools and methods for achieving these objectives.

More details on the linguistic process are presented alongside with the research approach in Sect. 3. After the linguistic processing, the technical processing is followed which follows the objectives of weighting the terms and transforming the data into a structure which is supported by the text mining methods/classical data mining methods which have to be applied in a further step (Schieber and Hilbert 2014). For the term weighting procedure, measurements metrics are required. These metrics can be classified in simple such as absolute, relative or document frequency and weighted metrics. The following metrics are often applied:

$TF \times IDF$ where TF is the term frequency and IDF is the inverse document frequency.

According to Schieber and Hilbert (2014) in addition to the described metrics, it is also useful introducing domain-specific weighting measures such as in Li and Liu (2012) or Beheshti-kashi et al. (2015), in particular, if the existing metrics does not deliver satisfactory accuracy.

The term weighting process enables additional advantages. One advantage is that the number of terms will be reduced by filtering the irrelevant words from the relevant words. Furthermore, the term weighting increases the accuracy of the results while conducting text mining methods. Chou et al. (2010) illustrate the different feature selection methods and compare them in their work (Schieber and Hilbert 2014).

The second objective of the technical processing is to transform the data into a format which is suitable for the application of text mining/data mining methods.

Table 1 Linguistic processing—sub-processes (based on Schieber and Hilbert 2014)

Sub-processes	Objective of sub-processes	Main tools/methods	Description of sub-processes
Word filtering	Filtering irrelevant word	Stop word lists	Filtering so-called stop words (articles, prepositions, etc.)
Lexical process	Reduction of terms to original forms	Lemmatization	Reduction to base form/dictionary form
		Stemming	Reduction to word stem
Syntactic process	Targeted extraction	Part-of-speech (POS) tagging	Assignment of POS to all terms
		Parsing	Identification of the syntax and the word functions of a sentence
Semantic process	Word meaning and relationships within the sentences	Calculation of collocations	Collocations are calculated identifying word combinations that occur often together
		Thesauri and ontologies	Usage of word databases such as WordNet (English), GermaNet (German)

The common data structure is the vector space model introduced from Salton et al. (1975). The different documents will be transformed into a vector and joined into a matrix. The organization of the matrix is as follows: the terms will be organized into columns, and the documents will be put as rows (figure Output 3). The output of the sub-process technical processing is the following:

1. Weighting and reducing of the terms quantity
2. Transformation of the terms into the vector space model

Without the technical processing, the application of text mining methods or classical data mining methods is not possible. After having transformed the text into a numerical format, the actual application of classical data mining algorithms such as classification or clustering follows. After the application of those algorithms, the results will be evaluated and the actions are derived.

In order to obtain a better understanding of the actual output of the different sub-processes, Fig. 3 displays a typical output of the tokenization, linguistic (POS tagging) and technical processes.

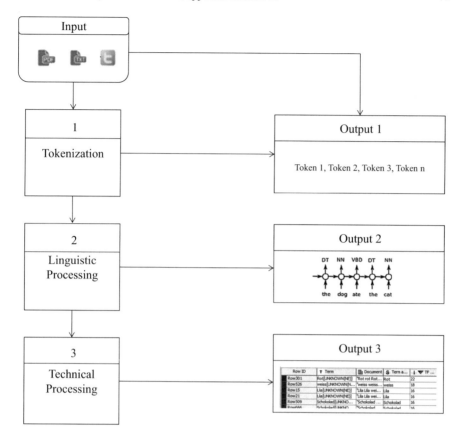

Fig. 3 Document processing steps output

3 Research Approach: Topic Detection and Tracking in Fashion Blogs

Corresponding to the emergence of social media, numerous research works have appeared which examine the different phenomena related to social media. Social media analytics (Fig. 4) as an interdisciplinary research field has arisen and integrates the different approaches. Structural attributes, opinion/sentiment-related and topic/trend-related approaches are summarized through the field of social media analytics (Stieglitz et al. 2014). The methods of these approaches are statistical analysis, social network analysis, sentiment analysis, content analysis and trend analysis.

Since fashion blogs are included, accordingly the research approach is based on the social media analytics framework. Within the framework, this research focusses topic/trend-related approaches, since the main interest is in the identification of fashion relevant content and their different characteristics and their development over

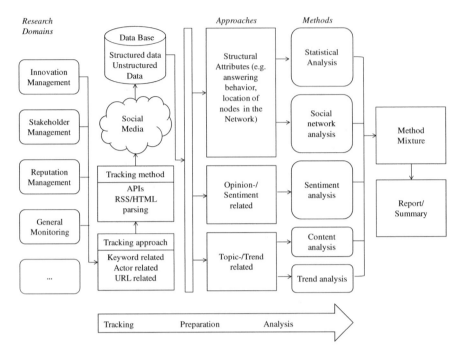

Fig. 4 Social media analytics framework (Stieglitz et al. 2014)

time. Figure 5 demonstrates the research approach of topic detection and tracking in fashion blogs.

Within topic/trend-related approaches, the present research approach is based on the principles of the topic detection and tracking (TDT) approach which origins from the field of intelligent information retrieval. TDT is a research programme and *evaluation paradigm addressing event-based organization of broadcast news* (Allan 2002). The objective of TDT is to identify individual news stories, in order to be able to monitor newly arriving events and to sort them into groups. For the tracking stream, already identified news stories are used as a baseline, to find further information on the topic (Allan 2002).

Since the objective of this research is to illustrate the different approaches on how to deal with the fashion blogs having in mind the buyer's benefits, the actual procedure of TDT is not followed in the present research. While the original TDT is considering both streams as a process, this research is dealing topic detection and tracking as separate tasks. The intention is more on the examination and illustration of the two different directions on how the blog posts might be applied.

A further difference is that this research is not focussing on events, but on features, more precisely on colours and topics. In the topic detection stream, the blog posts are the starting point of the analysis and the objective is to detect topics, which in the case of this research are colours, in order to make reveal what is inside the blog posts.

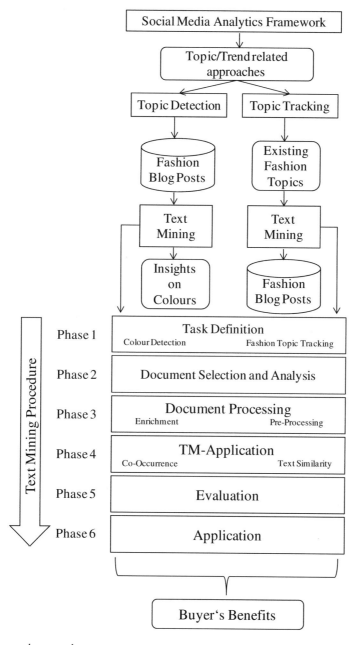

Fig. 5 Research approach

Table 2 Summary of the text mining steps

Phases and sub-processes	Topic detection	Topic tracking
Task definition	Detecting single colour occurrences	Tracking of existing fashion topics in the blog corpus
	Detecting co-occurred colour occurrences	
	Matching of colour occurrences on a timeline	
Document selection and analysis		
Source selection	74 German written fashion blogs	74 German written fashion blogs
Collection	TrendFashion tool	TrendFashion tool
Selected working environment	KNIME analytics platform	TrendFashion tool
Document processing		
	Tokenization	Tokenization
	Stop word filtering	Stop word filtering
	Stemming	Stemming
	Term frequency	POS tagging
		Customized metric
Text mining methods		
	Co-occurrence	Text similarity

In contrast, in the topic tracking stream, a particular topic extracted from fashion online magazines is provided as input data, and it is looked for in the blog posts for similarities, in order to examine if existing/real topics are mirrored also in the blog posts. Table 2 summarizes the different steps undertaken for the topic detection and topic tracking tasks.

Phase 1: Task Definition

The task definition is crucial since all of the following steps are dependent on the first phase. Hotho et al. (2005) define, for instance, the grouping of similar documents as one task.

For the topic detection stream, the task is to detect single and co-occurred colour occurrences in a first step. In a proceeding step, the detected occurrences are matched to a timeline. The timely assignment of the colours is crucial for the fashion scenario due to the high seasonality of the products and the industry. Therefore, with matching the colours on a certain time period, it is aimed to examine potential seasonal patterns. For the single colour occurrences, a 12-month time period (2016) was selected. In addition to the single occurrences of the colours, the co-occurrences of the different colours are relevant to be examined. Co-occurred colours are defined as colours that often occur together. This would allow us to obtain more valuable information from the corpus. Similarly, to the first approach, the findings are matched over a time period. However, in this case, the co-occurred colours are matched over a time period of 5 years. Accordingly, a dataset was generated from the fashion blog data considering the years 2012–2016. On this dataset, examinations on the extraction of colour and colour combinations (colours which occur often together) are conducted.

For the topic tracking stream, five different fashion topics from 2016 have been selected. The objective of this task is to examine the tracking possibility of fashion topics in fashion blogs. For this reason for the tracking, the task was conducted on the longer lasting time period of the last 5 years (2012–2016), and not only 12 months. The selected fashion topics are 60s, XXL Sleeves, Boho, Fringe and Metallic. These topics are only some examples of fashion topics, which were referred to, in several fashion magazines. They were selected because they represent different features which usually manipulate a fashion topic. For instance, Boho and 60s represent a style, fringe is a decorative application, metallic represents colours such as gold, silver and copper and XXL sleeves are representing the silhouette. With these examples, the tracking possibilities of different fashion manipulating features should be illustrated. These topics are extracted from the online magazines gofeminin, stylefruits and stylebook. The textual descriptions of those topics have worked as the baseline for the further processing and tracking.

Phase 2: Document Selection and Analysis of Source Documents

Phase 2 consists of three sub-processes: the first sub-process is the selection of the source documents: in this case, the selection of the fashion blogs. The second sub-process is determining the features of the source documents. The third sub-process is loading and storing the source documents in the actual working environment. As described in Sect. 2.1, fashion blogs are selected as source documents. For the selection of the blogs, different criteria have to be considered since the input data might have an impact on the final results. Within the last years, the number of fashion blogs has highly increased. Accordingly, the first question within this sub-process is which fashion blogs should be considered as source documents. One option would be to create an own ranking scheme, based on special ranking criteria such as the Alexa Rank or PageRank. However, the second option would be to take an existing ranking. For this purpose, an online ranking providing the 100 most influential fashion blogs written in German was selected as the baseline for the analyses. This list of 100 blogs was examined according to some further criteria such as the target group of the blog or the availability of the time period 2012–2016. According to this analysis, only fashion blogs focussed on women fashion, providing information on various products (for instance, not limiting only to sneakers) and posts available for the targeted time period 2012–2016 have been selected. Overall, 74 blogs are included in for further processing. Before the collection and loading of the blogs, it is required to examine and determine their features. Due to the highly seasonal character of fashion products, the collection of the actual date is essential for further analyses. In addition to the actual content and the date, the title and URL of the posts have been identified as the required features to be included in the dataset. In a further step, the blog posts have been collected using the TrendFashion tool (Beheshti-Kashi et al. 2015). The TrendFashion tool was designed for collecting, processing and storing fashion-related textual data from the Web in a database. The tool integrates the two main functionalities, data collection and data processing. For the data collection, a list of URLs is required as input. Within the data processing step, different natural language processing steps are conducted, such as tokenization, named entity recognition or POS tagging. For the purpose of our research, a list of blog URLs has been given as input, and blog posts have been collected and stored alongside with the title, date and the actual post. The output is stored in paragraph, sentence and word level.

For the topic tracking stream, the TrendFashion tool was used as working environment. For the topic detection stream, Konstanz Information Miner (KNIME) was used. KNIME was used in this research in particular since it is an open-source software and, secondly, because of its strong text processing nodes. Though the analyses may also be conducted in RapidMiner, or within a R or Python framework achieving similar results. In order to read the blog posts into KNIME, the posts are exported in XLS files from the TrendFashion tool database.

Phase 3: Document Processing

As aforementioned, the document processing phase is essential for the processing of textual data. This process consists of three sub-processes, namely term identification, linguistic process and technical process. The term identification, which is often called tokenization (Kit and Webster 1992), aims at splitting the free text into terms and was conducted in a first step in the current analysis. For this purpose, all punctuations are erased; paragraphs are segmented into sentences and sentences into words. In this context, also all words consisting of less than three chars and all the numbers are filtered out. This is followed by the linguistic analysis whose objective is to analyse the text from a language perspective. One advantage of this process is that the extraction of certain part-of-speeches or sentence functionalities is possible in a more targeted way. The linguistic analysis consists of the sub-processes: filtering stop words, lexical analysis, syntactic and semantic analysis (Schieber and Hilbert 2014). Within the process of stop word, filtering words such as articles, prepositions, conjunctions are filtered out. These words are mostly irrelevant for further analysis. In some cases, it might be useful to filter out also nouns or other words which might disturb the analysis.

In the lexical analysis, the remaining terms have to be analysed regarding their morphology. In order to do so, it is crucial to reduce all the terms to their original forms (Schieber and Hilbert 2014). This process can be conducted via two different methods: lemmatization and stemming. In the process of lemmatization, the terms are processed to their simple forms, namely for nouns into the first case, for verbs into the infinitive (Bird et al. 2009). Stemming is not the process of bringing back to the word origin, though to the reduction of the term itself (Bird et al. 2009). For the English language, the Porter stemmer is often applied (Porter 1980). Both methods can be used, though it depends on the analysis goal. The advantage of applying either lemmatization or stemming is the reduction of terms which will result in performance advantages. In the context of this work, stemming was applied as the main method within the lexical process. Since the corpus consists of German written posts, it was required to use a stemmer which is working for the German language. Therefore, the Snowball stemmer was selected since it is designed to stem languages such as German, French, Spanish and more.

The syntactic analysis follows the lexical analysis. During this process, mainly two sub-processes are conducted: part-of-speech (POS) tagging and parsing. Within the POS tagging, all terms have to be assigned to their corresponding POS (Bird et al. 2009). This is performed with the help of lexicons and probabilistic models (Bahl and Mercer 1976). The advantage of this process is that depending on the problem, the terms may be directly targeted. For the German language, the TreeTagger is used, which is based on a decision tree (Schieber and Hilbert 2014). For the English language, the Brill algorithm is usually applied (Bird et al. 2009). The parsing is focussed on the identification of the syntax and the word functions of a sentence. Therefore, at the end of this process information, for instance, on the subject or object of a sentence becomes evident. Since there was no interest in the functions of a sentence, the parsing step was not conducted in these analyses. The POS tagging step was only conducted for the topic tracking stream.

In the last step of the linguistic processing, the semantic level is focussed. For this purpose, it was required to conduct an enrichment step for the topic detection. For the colour-related analyses, a dictionary including colour names was generated, and the so-called dictionary tagger node in KNIME was used, to match the documents with the dictionary. The aim is to in the following step just filtering out those words with the colour tags. Using this approach for the enrichment, it is possible to directly target the colours and neglect the rest of the terms.

In order to apply further text mining methods, in the following step, it is required to apply the technical processing by weighting the terms and transforming in an adequate structure. For the topic detection task, simple term frequency has been applied, as measurement metric. For the topic tracking, a specified metric was developed and applied. This metric weighted the terms according to their relevance for the corresponding topic extracted from the textual descriptions from the online magazines. In the last step of the document processing phase, document vectors are created in order to transform the textual data into a numerical representation.

Phase 4: Application of Text Mining Methods

For the topic detection stream, co-occurrence analyses have been conducted in order to detect co-occurred colours. For the topic tracking stream, text similarity has been conducted, in order to find similarities between the existing topic descriptions and the actual coverage in the blog corpus.

Phase 5: Evaluation

The purpose of this step is to evaluate the results regarding the defined tasks and the quality of the results. For this, performance figures matching the tasks and the selected text mining methods are selected. For text classification, for instance, accuracy, precision, recall or the F-measure are used (Schieber and Hilbert 2014). For text similarity cosine similarity or the Jaccard coefficient may be applied (Subhashini and Kumar 2010).

Fig. 6 Number of posts and detected colour occurrences for a 12-month period

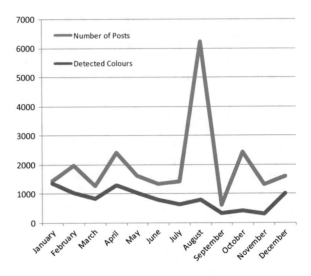

Phase 6: Application

From the results of Phase 3 and 4, the actual benefits for the buyer are discussed in Sect. 5.

4 Results on Experimental Analyses of Fashion Blogs

4.1 Topic Detection—Single Colour Occurrences

Figure 6 demonstrates the total number of posts for a 12-month period and the detected monthly colour occurrences in 2016. The total number of posts for the 12 months is 23,673, which is an average of 1972.75 posts per month. However, the number of posts for August is more than the triple of the average, which might be probably through the fashion shows during the summer and the corresponding coverage on the blogs. All the posts contain 9756 colour occurrences distributed in 145 different colours.

Table 3 shows the percentage of the detected colours over the 12 months. The top ten colours are white, black, red, grey, rose, blue, beige, gold, brown and green. One objective the analysis of the single colour occurrences was to examine whether the colours might be assigned to a certain season and to conclude some seasonal dependencies. However, from the analysis, it is not possible to conclude any seasonality of single colours within the fashion post corpus.

Table 3 Single colour trends for a 12-month period

	January	February	March	April	May	June	July	August	September	October	November	December
White	27.6	30.1	22.7	24.5	29.1	27.7	39.4	28.2	23.0	23.3	26.0	21.1
Black	12.6	11.6	9.6	10.0	10.6	10.6	8.5	8.3	7.9	8.7	11.3	10.9
Red	9.7	3.8	10.7	4.4	6.7	7.4	3.9	4.3	4.5	5.0	3.3	6.5
Grey	5.5	4.8	5.0	4.7	4.0	6.8	3.7	4.3	4.8	6.0	5.3	5.7
Rose	4.2	4.6	5.5	4.7	4.7	4.3	4.8	5.2	8.5	4.5	4.0	4.1
Blue	3.4	3.8	4.3	4.7	4.2	5.9	5.5	3.8	4.5	4.2	2.7	1.9
Beige	2.6	1.9	3.9	4.3	4.2	4.5	2.1	4.7	3.0	1.7	3.0	4.4
Gold	2.9	4.8	2.4	3.7	3.5	1.8	3.4	3.6	3.0	2.7	5.0	4.4
Brown	1.6	2.8	3.2	2.6	4.2	4.5	2.9	1.9	2.1	4.7	3.3	4.4
Green	2.6	3.0	2.8	1.6	1.9	2.3	3.2	4.2	5.1	2.5	3.0	3.2

4.2   Topic Detection—Co-occurred Colour Occurrences

In addition to the single colour occurrences for a 12-month period, further examinations on a fashion blog corpus, containing posts from the years 2012–2016, are
conducted. This corpus consists of 71,939 posts distributed over the 5 years.

The objective of the further analysis was to detect not only single colour occurrences but also co-occurred colours. In the post-distribution a yearly increase can be
stated, which might be due to increased relevance of the bloggers and their increased
activities within the last years. The number of blogs included in the analysis is the
same for the whole time period. Overall, 1800 co-occurred colour occurrences have
been detected. The objective of this analysis was to detect not only single occurrences
of colours but the detection of co-occurred colours. It is also notable that the colour
combinations do not refer to one article, but rather to a whole outfit.

Table 4 summarizes the top ten ranked colour combinations for 2012–2016.

The ranking is based on the corpus frequency of the ngrams. The corpus frequency
represents the frequency of the ngrams in the total dataset per year. The analysis
has shown that some colour ngrams (co-occurred colours) appear in each year, for
instance, black-white or red-white, and other ngrams only in particular years such
as apricot-chocolate or pink-purple. Furthermore, the analysis has demonstrated that
the ngrams and their frequencies respectively the ranks of the ngrams differ between
the years. In order to show the trend of the colour ngrams, and their different ranks
through the years, additional indications are included next to the colour ngrams. The
comparison is based on the rank of the previous year. In the case of no change, an
equal sign (=) is used and a completely new occurrence is illustrated with the asterisk
symbol (*). Else the increase or decrease is indicated by a $\pm$ and the corresponding
rank increase/decrease.

From the table, it is hardly possible to observe the development of the colour
combinations over the 5 years. Each year has mostly its own colour combinations.

4.3   Topic Tracking of Fashion Topics

For the topic tracking stream, we have used the same 5-year corpus, as for the topic
detection of co-occurred colours. Figure 7 displays the trends for the selected fashion
topics over the time period. Firstly, it is notable that is indeed possible to track the
fashion topics in fashion blogs. Secondly, the tracking results differ between the
different topics. For instance, for the fashion topics Boho and 60s we can observe
that they have an increased coverage in 2016 in the blogs, which correspond with the
fact that they are described as leading fashion topics for 2016. For the Fringes topic,
we can see that the topic has its peak in 2015 and is decreasing in 2016. In order to
understand this development, we have looked up fashion topics for 2015 in online
magazines and identified that it was also clearly discussed as a leading topic for 2015.
In addition, we have checked the Google Trends website, in order to compare the

Table 4 Ten most ranked colour ngrams and their development for 2012–2016

Rank	2012	2013		2014		2015		2016	
1	Black-white	Black-white	=	Black-white	=	Grey-rose	*	Black-white	+6
2	Rose-white	Pink-rose	*	Grey-white	*	Grey-white	=	Beige-light blue	*
3	Red-black	Orange-red	*	Grey-black	+4	Rose-red	*	Blue-light blue	*
4	Blue-yellow	Brown-black	*	Beige-grey	*	Rose-white	*	Light blue-white	*
5	Blue-red	Purple-pink	*	Blue-white	*	Yellow-orange	*	Light blue-rose	*
6	Red-white	Gold-silver	*	Blue-red	*	Black-white	−6	Blue-white	*
7	Apricot-chocolate	Grey-black	*	Grey-silver	*	Beige-rose	*	Grey-light blue	*
8	Yellow-red	Yellow-orange	*	Anthracite-grey	*	Rose-silver	*	Grey-white	−7
9	Grey-brown	Dark blue-white	*	Brown-grey	*	Beige-grey	−5	Yellow-white	*
10	Grey-white	Dark blue-black	*	Pink-black	*	Red-white	*	Beige-grey	−2

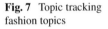

Fig. 7 Topic tracking
fashion topics

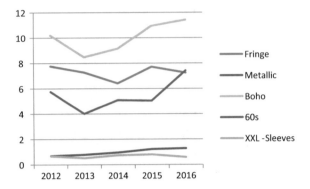

trend of the search keyword Fringe users with our results. Similarly, to our findings, the trend peaks in 2015 weaken in the course of 2016. Both additional findings might explain the tracking result of Fringe in our analysis. However, we have to mention that Google Trends only consider the keyword Fringe, whereas we have used textual descriptions of the fashion topic Fringe, processed it and finally worked with word lists describing the topic more detailed.

For the topic Metallic, we can observe a tiny increase in 2016. Considering the single colour occurrences, we can see that the colours grey and gold belong to the 12 colours in 2016. Both colours might refer to the topic Metallic. We have checked the description for the Metallic topic after the analysis and have realized that those colours are not included. In a typical text mining process, at this stage, some adjustments have to be performed and the process has to be repeated. The results for the XXL sleeves topic do not correspond with our primary assumptions.

From the analysis, no linear trend over the years is observable. However, we can conclude that the fashion topics weaken in some periods and will come back. In order to detect seasonal patterns, it is required to conduct further analysis on monthly or quarterly level.

5 Buyers Perspective—Discussion

This chapter has examined the detection and tracking of fashion topics in fashion blog posts. In a first step, we have focussed on the detection of single colour and co-occurred colour occurrences. In the second part of our examinations, different fashion topics have been tracked over a 5-year time period. It has been illustrated that it is possible to detect colour occurrences and to track existing fashion topics over a certain time period. However, the actual objective is not the examination of the detected content. The main goal is discussing the potential added value that a fashion buyer might gain from the described approach.

In order to examine potential benefits for the buyer, we need to make a step back and look at the status quo in the buyer's activities. As mentioned in the introduction section, the main goal of the buyer is to arrange a collection for a particular time period. For this task, he has to conduct amongst others trend monitoring activities before generating the actual production plans. Often these activities are impacted by subjective decisions and experiences of the buyers. This might turn into a high dependency of the company to individual buyers. This potential dependency might be weakened through the quantitative analysis of the fashion blogs. A typical scenario for a buyer is visiting fashion shows, which present next seasons' hottest fashion topics. Often these shows are 9 months before the actual market launching of the products. Based on these shows, companies often make adjustments in their productions orders (Teucke et al. 2014). Due to the fact that most production plants are located in Asian countries, and the actual target countries are European countries, the time to market of the products can take up to 6 months (Teucke et al. 2014). Following our analysis in this context, before finalizing the production plans, the buyer might have the possibility to extract information on colours, co-occurred colours or track topics presented on the fashion shows from the time between the shows and their order deadlines for the next collections. This procedure would allow him to catch the actual coverage on the different fashion topics from potential customers and consequently might reduce the uncertainty in the buyers own decisions of selecting the collection.

6   Conclusion and Outlook

This chapter has presented an approach on detecting and tracking colour occurrences and fashion topics on a fashion blog corpus. For this, some experimental analyses on the posts have been conducted, following a text mining procedure model. However, the main interest is not in the detection of certain colours, or the tracking of certain topics. The fashion topics are working as examples. The main objective is to derive from those exemplary analysis some potential benefits for the fashion buyer. The main added value that we conclude is that the buyer can quantify his own experiences, and individual perceptions on potential fashion topics of the seasons, through the described approach. The analysis has demonstrated that it is possible to extract single colour and co-occurred colour occurrences and to match them in a time period. In addition, it has shown that the topics are visible in the blogs, to a greater or lesser extent over the 5-year time period. This means that the blog posts contain sufficient fashion-related content which makes the tracking possible.

The next steps of the research are on different levels. Firstly, the analysis of the colours will be expanded. At the current status, the co-occurred colour occurrences are looked at on a yearly basis. However, these analyses have not shown any patterns throughout the years. The assumption is that it is required to conduct the analysis on a monthly basis, in order to identify some seasonal patterns related to the colours. Secondly, the results will be discussed with fashion buyers, in order to understand their actual needs and to adapt the analysis to them. Thirdly, a further goal is to examine the actual advance the colours and fashion topics can be detected and tracked on the blogs. For instance, it is possible to detect topics prior to their high emergence to the markets, and if it is possible, with which advance are we able to detect them. These results might reduce the uncertainty of the buyers' decisions for the actual collections selection.

References

Alby T (2007) Web 2.0. Konzepte, Anwendungen, Technologien. Carl Hanser Verlag
Allan J (2002) Topic detection and tracking event-based information organization. Springer Science & Business Media
Bahl LR, Mercer RL (1976) Part-of-speech assignments by a statistical decision algorithm. In: IEEE international symposium on information theory
Beheshti-Kashi S, Lütjen M, Stoever L, Thoben K (2015) TrendFashion—a framework for the identification of fashion trends. In: Cunningham I, Hofstedt D, Meer P, Schmitt K (eds) Informatik 2015, Lecture notes in informatics (LNI), Gesellschaft für Informatik. Bonn
Berry MW, Kogan J (2010) Text mining: applications and theory. Wiley
Bird S, Klein E, Loper E (2009) Natural language processing with python. O'Reilly Media
Bruns A (2006) Towards produsage: futures for user-led content production. In: Sudweeks F, Hrachovec H, Ess C (eds) Cultural attitudes towards communication and technology, 28 June–1 July, Tartu, Estonia
Chou C-H, Sinha A, Zhao H (2010) A hybrid attribute selection approach for text classification. J Assoc Inf Syst 11(9):491–518
Crawford Camiciottoli B, Ranfagni S, Guercini S (2014) Exploring brand associations: an innovative methodological approach. Eur J Mark 48(5/6):1092–1112
Halvorsen K, Hoffmann J, Coste-Manière I, Stankeviciute R (2013) Can fashion blogs function as a marketing tool to influence consumer behavior? Evidence from Norway. J Glob Fash Mark 4(3):211–224. https://doi.org/10.1080/20932685.2013.790707
Hotho A, Andreas N, Paaß G, Augustin S (2005) A brief survey of text mining. LDV Forum 20(1):19–62
Hu X, Liu H (2012) Text analytics in social media. In: Mining text data. Springer, pp 385–414
Jackson T, Shaw D (2001) Fashion buying and merchandising management. Palgrave
Kaplan AM, Haenlein M (2010) Users of the world, unite! The challenges and opportunities of social media. Bus Horiz 53(1):59–68. https://doi.org/10.1016/j.bushor.2009.09.003
Kit C, Webster J (1992) Tokenization as the initial phase in NLP. In: COLING '92 Proceedings of the 14th conference on computational linguistics—volume 4
Kulmala M, Mesiranta N, Tuominen P (2014) Organic and amplified eWOM in consumer fashion blogs
Li G, Liu F (2012) Application of clustering method on sentiment analysis. J Inf Sci 38(2):127–139
Miner G, Elder J, Fast A, Hill T, Nisbet T, Delen D (2012) Practical text mining and statistical analysis for non-structured text data applications. In: Miner G (ed). Elsevier Science Publishing Co, Inc
Porter MF (1980) An algorithm for suffix stripping. In: ACM SIGIR conference on conference on research and development in information retrieval
Power DJ, Phillips-Wren G (2012) Impact of social media and Web 2.0 on decision-making. J Decis Syst 20(3):249–261. https://doi.org/10.3166/jds.20.249-261
Rickmann TA, Cosenza R (2007) The changing digital dynamics of multichannel marketing. The feasibility of the weblog: textmining approach for fast fashion trending. J Fash Mark Manag 11(4)
Salton G, Wong A, Yang C (1975) A vector space model for automatic indexing. Commun ACM 18(11):613–620

Schieber A, Hilbert A (2014) Entwicklung eines generischen Vorgehensmodells für Text Mining. In: Dresdner Beiträge zur Wirtschaftsinformatik (2014), Nr. 69/14

Schmidt J (2009) Das neue Netz. Merkmale, Praktiken und Folgen des Web 2.0. UVK Verlagsgesellschaft

Shaw D, Koumbis D (2013) Fashion buying. From trend forecasting to shop floor. Bloomsbury

Stieglitz S, Dang-Xuan L, Bruns A, Neuberger C (2014) Social media analytics. Bus Inf Syst Eng 6(2):89–96. https://doi.org/10.1007/s12599-014-0315-7

Subhashini R, Kumar VJS (2010) Evaluating the performance of similarity measures used in document clustering and information retrieval. In: 2010 first international conference on integrated intelligent computing, pp 27–31. http://doi.org/10.1109/ICIIC.2010.42

Teucke M, Ait-alla A, El-berishy N, Beheshti-Kashi S, Lütjen M (2014) Forecasting of seasonal apparel products. In: Kotzab H, Pannek J, Thoben K-D (eds) Dynamics in logistics. Fourth international conference, LDIC 2014 Bremen, Germany, February 2014 proceedings. Springer

Wong WK (2013) Understanding key decision points in the apparel supply chain. Optimizing decision making in the apparel supply chain using artificial intelligence (AI). Woodhead Publishing Limited. http://doi.org/10.1533/9780857097842.1

Zerfaß A, Bogosyan J (2007) Blogstudie 2007. Inform ationssuche im Internet - Blogs als neues Recherchetool

Part II
AI for Textile Apparel Manufacturing and Supply Chain

Review of Artificial Intelligence Applications in Garment Manufacturing

Radhia Abd Jelil

Abstract Nowadays, apparel manufacturing enterprises are confronted with ever-increasing global competition and unpredictable demand fluctuations. These pressures compel manufacturers to continuously improve the performance of their production process in order to deliver the finished product within the most approximate period of time and the lowest production cost. However, consistent and optimal solutions are difficult to obtain under a fuzzy and dynamic manufacturing environment. Therefore, in response to the need for new approaches, a large (and continually increasing) number of efforts have sought to investigate and exploit the use of AI techniques in a variety of industrial applications. This chapter provides a systematic review of contemporary research articles related to the application of AI techniques in garment manufacturing. The research issues are classified into three categories, including production planning, control, and scheduling; garment quality control and inspection; and garment quality evaluation. The challenges facing adoption of AI technologies in garment industry are discussed.

Keywords Artificial intelligence · Garment manufacturing · Decision making
Survey

1 Introduction

The garment industry is one of the most important sectors of the economy in terms of investment, revenue, trade, and employment generation all over the world. It is highly segmented and produces a wide variety of clothing and fashion products that change frequently with changes in style and season. The garment manufacturing involves many processing steps, beginning with order receiving and ending with dispatching shipment of the finished garments. Based on the present apparel industry,

R. Abd Jelil (✉)
Textile Materials and Processes Research Unit MPTex, Higher Institute of Fashion Crafts of Monastir, 5000 Monastir, Tunisia
e-mail: abdjelilradhia@yahoo.fr

© Springer Nature Singapore Pte Ltd. 2018
S. Thomassey and X. Zeng (eds.), *Artificial Intelligence for Fashion Industry in the Big Data Era*, Springer Series in Fashion Business,
https://doi.org/10.1007/978-981-13-0080-6_6

these processing steps can be categorized as pre-production, production, and post-production processes. The pre-production processes include sampling, sourcing of raw material, cost analysis, and approving the proposed product. The production processes include cutting and sewing. The postproduction processes include thread trimming, pressing, checking, folding, packing, and shipment inspection. Each step of these processes has its own set of considerations and requirements that should be addressed and completed before moving to the next phase. This makes the whole apparel manufacturing process incredibly complex and overwhelming to understand and manage.

Indeed, apparel companies have to deal with a global and very competitive environment which is characterized by rapid changes, short life-cycle products, high volatility, low predictability, tremendous product varieties, short production lead times, increasing customer demand, and rising labor costs. In such circumstances, industrial manufacturers constantly face complex and critical decisions. Traditionally, statistical regression methods have been widely used to solve decision problems, since they are easy to develop and implemented. However, they have the limitations of that they cannot handle complex nonlinear relationships with so many variables. Therefore, artificial intelligence (AI) techniques have been proposed as an alternative approach for modeling such complex relationships.

AI is the branch of computer science that is concerned with making computers behave like human beings. It deals with intelligent behavior which involves learning, reasoning, perception, communication, and interaction with complex environments (Adeli 2003). Many tools are used in AI including artificial neural networks (ANN), genetic algorithms (GA), fuzzy set theory, expert systems, machine learning. In the recent years, these techniques have attracted much attention of researchers and practitioners in the apparel manufacturing industry and have been applied successfully to solve a wide variety of decision-making problems such as production planning (Wong et al. 2014; Guo et al. 2013), cut-order planning (Bouziri and M'hallah 2007; Wong and Leung 2008), marker making (Huang 2013; Ozel and Kayar 2008), line balancing (Chen et al. 2002; Guo et al. 2006), sewing automation (Silva et al. 2004; Carvalho et al. 2010), inspection decisions (Zhang et al. 2011). Numerous research studies have shown that AI techniques have the potential of providing superior solutions over classical approaches (Guo et al. 2011).

This chapter aims to review current applications of artificial intelligence in garment manufacturing over last two decades. Based on the literature reviews, the challenges encountered by AI techniques used in the clothing industry will be discussed. The remainder of this review is structured as follows. The forthcoming three sections include review of applications of AI techniques in production planning, control, and scheduling, in garment quality inspection, and in garment quality evaluation. Afterward, challenges facing adoption of AI techniques in the garment industry will be discussed in the fifth section. Finally, conclusions will be given in the last section.

2 Applications of AI to Production Planning, Control, and Scheduling

Production planning, control, and scheduling is one of the most important aspects of garment industry which plays a vital role in coordination of the flow of materials and information between customers and suppliers and the business determining the product value stream. It manages the flow of material, the utilization of employee and equipment, and it responds to customer expectations. However, this process is prone to numerous disturbances which make it difficult to attain effectiveness of the actions taken. Therefore, the application of artificial intelligence techniques in all areas of production planning and control will make it possible to manage the knowledge in the area and enable improvement in the degree to which the customers' expectations will be met.

In this section, we will present a selection of the most significant applications of artificial intelligence techniques in some areas of the production planning and scheduling domain, including production order scheduling, cut-order planning, marker making, spreading and cutting schedules, line balancing, and machine layout design.

2.1 Production Order Scheduling

In the fashion industry, order scheduling is a key decision-making process which focuses on the assignment of production orders to appropriate production lines. However, in the most real-world apparel manufacturing environments, this process can be frequently disrupted through production difficulties and absenteeism. As a result, the pre-established order schedules are shifted very often after the production starts, which may lead to decreased production efficiency. Therefore, robust optimization approaches for production scheduling have gained increasing attention. In this way, Tang et al. (2017) investigated robust order scheduling problems in the fashion industry with the aid of a multi-objective evolutionary algorithm called non-dominated sorting adaptive differential evolution (NSJADE), by taking into account the pre-production events and the uncertainties in the daily production quantity. The NSJADE was utilized to search the order schedules in the fashion industry that achieve the following three objectives: (1) the schedules can minimize the total pre-production event clashes of all orders; (2) the schedules can minimize the total tardiness of all orders; (3) the schedules are not sensitive to variation of the daily production quantity during the process of real production. The observation from the experiments showed that the pre-production events and the existence of uncertainties in the daily production quantity heavily affect the order scheduling. Also, it was found that robust order schedules can be shifted less often after the production starts than non-robust ones, which saves labor cost and enhances the production efficiency. The authors underlined that with the help of robust order schedules, planners can pay close attention to the unfinished pre-production events as early as possible, negotiate earlier with the customers who place the orders about the delay in delivery, or arrange operators to work extra hours for these orders. In the same way, Guo et al. (2013) inspected a multi-objective order allocation planning problem in make-to-order manufacturing with the consideration of various real-world production features. To tackle this problem, they developed a novel hybrid intelligent optimization model, integrating a multi-objective memetic optimization process, a Monte Carlo simulation technique, and a heuristic pruning technique. The experimental results showed that the proposed model can effectively solve the investigated problem by providing effective production decision-making solutions. Wong et al. (2014) investigated a real-world production planning problem, multi-objective-order allocation, with the consideration of multiple plants and multiple production departments. They developed an intelligent and real-time multi-objective decision-making model to provide timely and effective solutions for this problem, by integrating RFID technology with intelligent optimization techniques. In this model, the RFID technology was used to collect real-time production data. Furthermore, a novel $(\mu/\rho + \lambda)$-evolution strategy process with self-adaptive population size and novel recombination operation was proposed and integrated with effective non-dominated sorting and pruning techniques to generate Pareto optimal production planning solutions. The obtained results showed that the proposed model can effectively solve the investigated problem by providing production planning solutions superior to industrial solutions.

Wong and Chan (2001) proposed an effective genetic algorithm approach incorporated with "earliness" and "tardiness" production scheduling and planning method to plan the clothing manufacturing process. In addition, a segmentation strategy was developed to divide the production planning period to overcome the problem of chromosome selection in GA. The experimental results demonstrated the effectiveness of the proposed method in the clothing manufacturing process. Mok et al. (2013) developed intelligent apparel production planning algorithms that allocate job orders to suitable sewing units to ensure the effective utilization of production capacity and on-time completion of all job orders. The intelligent planning algorithms were based on group technology and genetic algorithms and can be used for labor-intensive operation planning. The proposed algorithms have been shown to be able to substantially improve planning quality. The authors pointed out that these planning algorithms are currently used by apparel manufacturers in Hong Kong as part of their routine planning operations.

2.2   Cut-Order Planning

Cut-order planning (COP) is the problem of planning the fabric cut for a set of apparel orders. COP occurs for each order to be produced and is the starting point in the manufacture of the order. It seeks to minimize the total manufacturing costs by developing feasible cutting order plans with respect to material, machine, and labor. However, the COP process is a dynamic function that must respond to ever-changing status of many critical factors such as sales, inventory levels, raw materials, and labor and equipment availability. The variety of sizes, styles, fabrics, and colors induces significant complexity in this problem (Cooklin et al. 2006). Thus, effective optimization of COP solutions requires the use of advanced techniques derived from artificial intelligence. In this way, Wong and Leung (2008) proposed a genetic optimized decision-making model using adaptive evolutionary strategies in order to assist the production management of the apparel industry in the decision-making process of cut-order planning. The experimental results showed that the proposed method can reduce both the material costs and the production of additional garments while satisfying time constraints set by downstream sewing department. Bouziri and M'hallah (2007) investigated the cut-order planning problem using a new hybrid heuristic called "genetic annealing." This heuristic combined the advantages of population-based approaches (genetic algorithms) with those of local search (simulated annealing). The stopping criterion of this combined algorithm is to stop if the best current solution is not improved for three consecutive iterations. The obtained results demonstrated the validity and effectiveness of the proposed approach.

Abeysooriya and Fernando (2012a) presented a canonical genetic algorithm approach to solve the problem of cut-order planning generation. The proposed GA was implemented to maximize the number of garments in the cut templates generated in the COP, by searching optimized size ratios of the cut templates. Experimental results indicated that the proposed method can yield better solutions compared to the available methodologies of generating cut-order plans available in apparel industry. In another study (Abeysooriya and Fernando 2012b), the authors highlighted that adjoining conventional heuristic approaches with genetic algorithm accomplished an efficient searching of cut-order planning solutions.

2.3 Marker Making

Marker making is a critical process in the fabric-cutting room, in which pattern pieces of different sizes and styles of a garment are laid out on a sheet of paper, known as the marker paper, with fixed width and arbitrary length in order to achieve the highest fabric utilization (marker efficiency). One cutting order may require several markers to achieve optimal efficiency. Marker efficiency is determined by the percentage of the total fabric that is actually used in garment parts and depends on how tightly the pattern pieces fit together within the marker. The area in between the pattern pieces, which is not used by garment parts, is waste. The minimization of this waste is crucial to the reduction of production costs (Dumishllari and Guxho 2016). In fact, higher material utilization is of particular interest to garment industry, since a small percentage improvement in fabric efficiency can result in many savings. Thus, in order to optimize fabric utilization and achieve higher marker efficiency, increased attention is directed toward meta-heuristic methods, such as GA and particle swarm optimization algorithm, which are able to search the space searching intelligently (Hopper and Turton 2001). In this way, Vorasitchai and Madarasmi

(2003) investigated the pattern layout optimization problem using genetic algorithm in order to minimize fabric wastage in garment production industry. It was found that the proposed algorithm can improve the efficiency of almost all production quality markers, shirts, trousers, and other garments. Huang (2013) reported that optimized particle swarm algorithm can achieve ideal material utilization. M'hallah and Bouziri (2016) combined COP and marker making into a single problem, which was solved using constructive heuristics and three meta-heuristics: a stochastic local improvement method, global improvement method, and hybrid approach. These approaches were designed to enclose the behavior of expert markers and to limit the computation time. The obtained results provided computational proof of the benefits that industry can rip by integrating COP problem with marker making problem.

Wong and Leung (2009) proposed a methodology that hybridizes a heuristic packing (HP) approach based on grid approximation with an integer representation-based $(\mu + \lambda)$ evolutionary strategy (ES) in order to obtain an efficient layout of garment patterns so as to optimize the fabric utilization. The grid approximation provided two advantages over the geometric representation: The first one is that there is no need to introduce additional routines in order to identify enclosed areas in patterns, and the second is that it is easier to detect any overlap. The results showed that the proposed methodology provides an effective means by which to increase the marker efficiency. In another study, Wong et al. (2013a) developed a packing approach that integrates a grid approximation-based representation, a learning vector quantization neural network, a heuristic placement strategy, and an integer representation-based $(\mu + \lambda)$-evolutionary strategy to obtain efficient placement of irregular objects. The results were compared with those obtained by a genetic algorithm-based packing approach and those generated from industrial practice, demonstrating the effectiveness of the proposed approach.

Estimating the optimum cutting time directly affects the enterprise's cost, profit, product competition ability, and economic benefits. Ozel and Kayar (2008) designed a multilayer perceptron neural network to estimate the marker making cutting time based on marking lengths, the number of fabric layers, cutting blade speed, number of sizes, marking lengths, and cutting time. The network training was performed using the error back-propagation algorithm. It was found that the designed network exhibited satisfactory performance.

Hence, it can be drawn that the optimization of marker making process using artificial intelligence techniques can not only reduce the staff's working strength, but also greatly improve the material utilization ratio and the producing speed, leading to considerable economic benefits to garment manufacturing enterprises.

2.4 Fabric Spreading and Cutting Schedules

Fabric spreading is a preparatory operation for cutting that consists of laying plies of fabric one on top of the other in a predetermined direction on the cutting table to form a fabric lay. The composition of each spread, i.e., the number of plies for each color, is obtained from cut-order plan. Once spreading is done, garment panels are needed to be cut. Marker prepared according to the cut-order plan is plotted on a marker sheet, and it is laid on top of the fabric lay, to act as the reference for cutting. Cutter should follow the outline of the panels plotted on the marker sheet to cut the required cut panels from the fabric lay. Cutting out the patterns through all plies creates a set of bundles of garment pieces, and several such lays may be required to satisfy all demands (Rose and Shier 2007).

Scheduling of spreading and cutting demands labor cost minimization, faster throughput, greater accuracy, higher fabric utilization and correct cut-piece fulfillment. The problem consists of determining the lowest cost spreading and cutting schedule for garments of different styles, colors, and sizes, subjected to physical constraints of cutting table length and cutting knife height as well as business constraints of required demand for each stock-keeping unit (SKU) (Nascimento et al. 2010). In most cases, planning and scheduling of cutting process is decided by the managers based on the experience they gained by handling previous orders, so the systematic and effective functioning of the process is diminished. Furthermore, this subjective nature would not guarantee the optimal planning and scheduling of the process (Wong 2003). Moreover, in apparel manufacturing process, some dynamic factors which occur internally and externally will make the schedule complicated creating a harder problem to solve (Wong et al. 2005a, b). Therefore, effective scheduling is essential to accommodate higher production performances and low production costs. Recent advances in computing technology, especially in the area of computational intelligence, can be used to handle this problem. Among the different computational techniques, genetic algorithms are particularly suitable. A major feature of GA is the ability to take care of a variety of objective functions. Patrick et al. (2000) reported that optimal roll planning can be worked out by using genetic algorithm approach. They underlined that it is possible to save a considerable amount of fabric when the best roll planning is used for the production. Wong et al. (2005a) proposed a genetic algorithms approach to optimize both the cut-piece requirements and the makespan of the conventional fabric-cutting departments using manual spreading and cutting methods. It was found that both the makespan and cut-piece fulfillment rates were improved and that the latter was improved significantly. In another paper, Wong et al. (2005b) showed that the makespan and the influence caused by the change of schedule could be minimized by using a real-time GA based segmentation rescheduling approach.

Wong et al. (2006a, b) used genetic algorithms and fuzzy set theory to generate just-in-time fabric-cutting schedules in a dynamic and fuzzy environment. It was found that the genetically optimized schedules improved the internal satisfaction of downstream production departments and reduced the production costs simultaneously. Mok et al. (2007) proposed a fuzzification scheme to fuzzify the static standard time so as to incorporate some uncertainties, in terms of both job-specific and human-related factors, into the fabric-cutting scheduling problem. They also proposed a genetic optimization procedure to search for fault-tolerant schedules using genetic algorithms, such that makespan and scheduling uncertainties were minimized. Experimental results indicated that the genetically optimized fault-tolerant schedules not only improved the operation performance but also minimized the scheduling risks.

2.5 Assembly-Line Balancing

An assembly line consists of a number of workstations which are arranged along a material handling system, in order to obtain a sequence of finished product types. The work pieces are moved from station to station, and at each one, certain operations are performed in view of some constraints. The first primary constraint is the cycle time which corresponds to the maximum available time for the production of any work piece at any workstation. In addition to cycle time, precedence relationships, which specify the order in which tasks must be performed in the assembly process, are the other primary constraints (Eryuruk et al. 2008). Moreover, tasks are assigned to operators depending on the constraints of different labor skill levels. Inappropriate workstations assignment will lead to the increase of labor cost, work in process (WIP), cycle time, and poor throughput.

The assembly-line balancing (ALB) problem consists of assigning tasks to an ordered sequence of workstations so that each workstation has no more than can be done in the workstation cycle time, and so that the unassigned (that is, idle) time across all workstations is minimized. Task allocation is based on the objective of minimizing the workflow among the operators, reducing the throughput time as well as the work in progress, and thus increasing the productivity. Each operator then carries out operations properly, and the work flow is synchronized. ALB problems that occur in real-world situations are dynamic and are fraught with various sources of uncertainties such as the performance of workers and the breakdown of machinery. Thus, several investigations using soft computing methods have been carried out in an effort to improve the productivity and efficiency of assembly lines in the clothing industry. Among these techniques, GA method has received much attention and has been applied successfully to many optimization problems. One of the earliest applications of GA to the clothing industry was carried out by Chan et al. (1998), who applied their method to an assembly system manufacturing men's shirts. The authors tried to improve the line efficiency by minimizing the time spent in assembly-line balance planning. They also included the various skill levels of workers as problem-specific information to solve a 41-task ALB problem. The results showed that the performance of genetic algorithm was much better than the performance of the greedy algorithm, which performed optimization by proceeding to a series of alternatives and assigned the most skillful worker to each task. Chen et al. (2002) presented a hybrid genetic algorithm approach for assembly-line planning problems involving various objectives, such as minimizing cycle time, maximizing workload smoothness, minimizing the frequency of tool change, minimizing the number of tools and machines used, and minimizing the complexity of assembly sequences. They classified the assembly-line planning problems into line balancing, tooling, and scheduling problems. The proposed method was improved by including heuristic solutions into initial population and developing a self-tuning method to correct infeasible chromosomes. Experimental results indicated that the proposed method can efficiently yield many alternative assembly plans to support the design and operation of a flexible assembly system. Guo et al. (2006) developed a genetic optimization method capable of dealing with a garment assembly line producing multiple products. Its performance was verified through experimentation using empirical data. Wong et al. (2006a, b, 2013b) developed a line balancing technique using genetic algorithms for optimizing the assignment of operatives in an assembly line. They also investigated the impact of different levels of skill inventory on the assembly makespan in order to find out the optimal number of task skills an operative should possess in the apparel assembly process. In a practical case study application, the algorithm was shown to be efficient. Results also indicated that there was a margin of diminishing returns in terms of worker training, in that workers who could perform more than three sewing operations brought little benefit in terms of line balance (Wong et al. 2006a, b). Yolmeh and Kianfar (2012) designed an efficient GA to solve setup assembly-line balancing problem. To determine the assignment of tasks to stations, the algorithm was hybridized using a dynamic programming procedure. Using dynamic programming, at any time a chromosome could be converted to an optimal solution. The computational results showed that the proposed GA outperforms all of the algorithms presented to solve assembly-line balancing problems so far. Unal et al. (2009) developed a heuristic algorithm for line balancing and evaluated its effectiveness under different line configurations using simulation. It was found that U-type line configuration is more advantageous compared to straight-line configuration according to both mean throughput per worker and also mean workstation utilization values. The authors highlighted that the proposed simulation-based line balancing approach can be used in all types of garment production.

The above-mentioned publications illustrate the potential of GAs to address the garment industry assembly-line balancing problems. However, one limitation of GA is that it cannot easily incorporate problem-specific information. If better solutions can be achieved by including this type of information, it may be advantageous to enhance GA in this manner. Brown and Sumichrast (2005) indicated that grouping genetic algorithms (GGA), firstly proposed by Falkenauer (1993), were more efficient for solving grouping problems than the standard GA. Hence, some researchers investigate the use of these tools to handle ALB problems in garment manufacturing. Chen et al. (2012) developed a GGA to solve the ALB problem of sewing lines with different labor skill levels in garment industry. The developed GGA could allocate workload among machines as evenly as possible for different labor skill levels, so the mean absolute deviations (MAD) could be minimized. The computational results revealed that GGA outperformed GA in both simple and complex problems by 13.81% and 8.81%, respectively (Chen et al. 2009). The authors highlighted that production managers could use the research results to quickly design sewing lines and achieve higher labor utilization rates and higher throughput levels. Chen et al. (2014) employed GGA in order to solve the line balancing problem in terms of minimizing the number of workstations for a given cycle time (the type I ALB problem). The efficiency of the developed heuristic was verified using empirical data from a sportswear factory combined with computational experiments. The authors stated that their arrived at algorithm could be of great aid to production managers interested in reducing cycle times and increasing labor utilization.

2.6 Machine Layout Design

Machine layout or flow line design involves determining the relative positions of machines (i.e., the layout) in facilities where a given product is manufactured. The layout design is key concern to organize operations in such a way as to maximize resource usage and overall system throughput. It generally depends on the products' variety and the production volumes (Islam et al. 2014). Indeed, change of machine layout is often required for small quantity and diversified orders in the apparel manufacturing industry. Hence, facility layout design is a continuous iterative process based upon the changing constraints of dynamic environment (Naik and Kallurkar 2016). Poor layout can lead to inefficiency, inflexibility, large volumes of inventory and work in progress, high costs, and unhappy customers. Changing a layout can be expensive and difficult, so it is best to get it right first time. Thus, optimization of layout design has become very essential to improve operational efficiency and reduce the nonproductive time. Thereby, many efforts have been undertaken to develop new design models and procedures that account for uncertainty and variability in design parameters such as product mix, production volumes, and product life cycles, for complex manufacturing system analysis and rational decision making while handling.

Genetic algorithms have proven to be a valuable method for solving a variety of hard combinatorial optimization problems. Martens (2004) developed a pair of GAs based on two alternative integer programming (IP) models in an attempt to find quality solutions for a variety of large real-life layout cases in the fashion industry. The obtained results showed that the proposed GAs were able to find optimal or near-optimal solutions on small problem instances and that they were capable of solving large, real-life layout problem in the fashion industry in an acceptable amount of time. Lin (2009) addressed a single-row machine layout problem with the objective of minimizing moving distance for cut pieces in a U-shaped sewing line. They developed a hierarchical order-based genetic algorithm, which has been shown to be able to make random and global searches to determine the optimal solution for multiple sites simultaneously and also to find speedily the best order for machine layout, which shortens the pieces moving distance and enhances production efficiency. Ultutas and Islier (2015) developed a clonal selection-based algorithm to solve the dynamic facility layout problem in footwear industry. Several scenarios were generated by using the real-life data. The proposed algorithm has been implemented and tested, showing promising results.

From this analysis, it appears that accurate machine setup is a key factor for increasing productivity and that facility layout remains an open research issue.

3 Garment Quality Control and Inspection

Garment inspection is an important stage of quality control that still relies heavily on trained and experienced personnel checking semifinished and finished garments visually. This process alone is very time-consuming because of the variety of styles, sizes, and fabric used in the clothing. Another concern is the quality standardization which requires the inspection to be repetitive to achieve a certain satisfaction (Fung et al. 2011). However, manual inspection imposes limitations on identifying defects in terms of accuracy, consistency, and efficiency, as workers are subject to fatigue or boredom, and thus inaccurate, uncertain, and biased inspection results are often produced. To tackle these problems, it is necessary to set up an advanced inspection system for garment checking that can decrease or even eliminate the demand for manual inspection and increase product quality. In this way, artificial intelligence techniques can be explored to ensure reliable and accurate quality control in industrial apparel manufacturing.

In this section, we will present a survey of the main applications of AI techniques for quality control in garment manufacturing process.

3.1 Seam and Fabric Sewing Performance

In cut and sewn apparel products, seams are formed when two or more pieces of fabric are held together with stitches. Various seams can be obtained by combining different fabric-cutting, joining, and stitching parameters (Yildiz et al. 2013). As the seam is one of the basic requirements in the construction of apparel, seam quality has great significance in apparel products. The seam performance is affected by various fabric mechanical properties with a combination of their sewing parameters. Therefore, investigating the performance parameters' relations will help to get a better understanding of the sewing process.

Owing to the complex and nonlinear relationships between the above-mentioned factors, ANNs present an attractive alternative to conventional statistical predictive techniques. In this way, Gong and Chen (1999) have successfully used ANN to predict the making-up performance of fabrics during garment manufacture, based on their mechanical properties measured by the Kawabata Evaluation System (KES-F) system. The predicted making-up performances included laying-up, cutting, overall handling, inter-ply shifting, structural jamming, seam slippage, needle damage, seam pucker, ease of pressing, dimensional stability, and appearance retention. The authors highlighted that artificial neural networks are effective for predicting potential problems in clothing manufacturing. Patrick et al. (2007) predicted the sewing performance of fabrics from fabric composition, weave structure, yarn count, sewing thread properties, and fabric low-stress mechanical properties, using a three-layer back-propagation (BP) neural network, which consists of 21 input nodes, 21 hidden nodes, and 16 output nodes. The sewing performance included severity of seam pucker, severity of needle damage, distortion, and overfeeding. The predicted values of most fabrics were found to be in good agreement with the results of sewing tests carried out by domain experts. Hui and Ng (2005) investigated the use of extended normalized radial basis function (ENRBF) neural networks compared to traditional BP to predict the sewing performance of fabrics in apparel manufacturing. It was found that the ENRBF neural network had better predictability than the BP neural network and that both models provided better advice than the experts in some areas, when compared to actual sewing performance. In another study, Hui and Ng (2009) used an ANN technique based on a back-propagation algorithm with weight decay technique and multiple regression with common logarithm method to predict the seam performance of fifty commercial woven fabrics used for the manufacture of men's and women's outerwear based on seam puckering, seam flotation, and seam efficiency. The developed models were assessed by verifying mean square error (MSE) and correlation coefficient (R-value) of test data prediction. The results indicated that the ANN model performed better than multiple regressions and that the prediction errors of ANN were low despite the availability of only a small training data set. Thus, it can be concluded that ANN techniques could be emulated as human decision in the prediction of sewing performance of fabrics more effectively.

The prediction of seam strength is also very important because it affects both the functional and aesthetic performance of an apparel product in terms of durability and stability. Onal et al. (2009) studied the effect of fabric width, folding length of joint, seam design, and seam type on seam strength of webbings made from polyamide 6.6 which were used in parachute assemblies as reinforcing units for providing strength by using both Taguchi's design of experiment (TDOE) as well as an ANN. It was found that the predictions given by ANN model were better in accuracy than those performed by TDOE. Yildiz et al. (2013) constructed multilayer perceptron and radial basis function (RBF) neural network models to predict the seam strength and elongation at break in poplin and gabardine woven fabrics based on stitch type, seam density, sewing needle type, and sewing yarn type. The experimental results showed that both models produced reliable estimates of seam strength and elongation at break. The authors underlined that with the help of ANN models, sewing parameters can be chosen to form an optimum sewing process, leading to cost and lead time reduction.

3.2 Sewing Automation Equipment

The knowledge and control of sewing machines is an important aspect to consider by apparel manufacturers in order to produce high-quality garments and improve production efficiency. However, due to the complexity of the sewing process itself, commercial machines are not yet fully controlled or monitored. The sewing parameters are still being adjusted by "trial and error" at the beginning of the operation, as average values (Silva et al. 2004). Therefore, automation equipment is the key to improving the quality of apparel products and enhances the critical machine functions. Thus, many research efforts have been undertaken to avoid empirical machine settings, reduce setup times, and improve sewing machine performance and flexibility. Barrett et al. (1996) developed a wavelet neural network-based online classifier of fabric type and number of plies for use on a sewing machine to improve stitch formation and seam quality. Needle penetration forces and presser foot forces were captured and decomposed using the wavelet transform. The wavelet-filtered needle force waveform was used as an input of the ANN. It was found that the wavelet ANN could correctly classify both fabric type and number of plies being sewn with 97.6% accuracy. The authors indicated that, given the ability to identify fabric/ply combinations online, the sewing machine can use predefined sewing parameters to automatically adjust the sewing machine settings. Carvalho et al. (2010) developed a combined proportional integral derivative (PID)/fuzzy logic controller to provide a reference and force offset adaptation to the number of plies and sewing speed. The proposed controller included a "teach-in" procedure to tune the controller's parameters while varying sewing speed and the force output independently. The authors highlighted that the control system is still somewhat limited due to the dynamic response of the actuator and that future developments in the field of actuators will certainly allow a further improvement of the feeding behavior. They also stated that their work is a very significant contribution to eliminate trial-and-error tuning of the sewing machine, toward a highly flexible and controlled operation. Guhr et al. (2004) designed a fuzzy logic controller to control the vertical movement of the presser foot of the overlock sewing machine, which has been shown to effectively improve the control performance by adapting the sewing parameters such as speed, number of plies, and type of fabric.

Koustoumpardis and Aspragathos (2003) proposed a hierarchical robot control system which includes a fuzzy decision mechanism combined with a neuro-controller to regulate the tensional force applied to the fabric during the robotized sewing process. The fuzzy logic decision mechanism utilized only qualitative knowledge concerning the properties of the fabrics, in order to determine the desired tensional force and the location of the robot hand on the fabric. A feed-forward NN controller regulated the fabric tension to achieve the desired value by determining the robot end effector velocity. The simulation results demonstrated the efficiency of the system as well as the robustness of the controller performance since the effects of the noise are negligible. In another study (Koustoumpardis and Aspragathos 2007), the authors implemented and tested the proposed system in a real robotized sewing environment for two fabric handling tasks: firstly for a robot guiding fabrics toward sewing and secondly for the cooperation of a robot with a human for handling fabrics. It was found that the robot demonstrated a satisfactory real-time response and that the neural network controller was more robust than a PID controller working under identical conditions. Zacharia et al. (2009) designed a robot control system based on visual servoing and fuzzy logic for handling fabrics lying on a work table. This system was enhanced using genetic-based and adaptive control. The experiments showed that the proposed robotic system was flexible enough to handle various fabrics and that it was robust in handling deformations that may change fabric's shape due to buckling (wrinkling and folding). The authors indicated that this framework does not cover all the aspects of robot handling of flexible materials, since there are still several related issues requiring solutions. To alleviate the computational burden of geometrical computations, Zacharia (2012) proposed an innovative approach based on a novel genetic-oriented clustering method and an adaptive neuro-fuzzy inference system (ANFIS) for robot handling pieces of fabrics with curved edges toward sewing. The experimental results showed that the proposed approach was effective and efficient in guiding the fabric toward the sewing needle, sewing it, and rotating it around the needle and that it was robust against fabric's deformations. Also, it was shown that the proposed method presented good results when applied to fabrics with curved edges of unknown curvatures. The authors highlighted that this approach is applicable to any piece of fabric with edges of arbitrary curvature and that the achieved accuracy is really promising for future use in industrial applications

Kim et al. (2004) constructed a nonlinear network model for a commercial sewing machine equipped with a brushless direct current (BLDC) motor. Based on the model, a two degrees-of-freedom (DOF) PID controller was designed to compensate the effects of disturbance without degrading tracking performance. According to the experimental results, the model has been shown to be a good approximation of the sewing machine and the proposed method demonstrated the effectiveness for a motion control system that requires high speed, robustness, and accuracy.

Fung et al. (2011) developed a novel and flexible 6-axis robotic hanger system with three DOFs which can move in the two-dimensional (2D) plane for the inspection of knitted garments with different styles, sizes, and cloth fabrics. This 3-DOF hanger consists of three groups of linkages (body link, shoulder link, and sleeve link) which were designed to satisfy the conditions of inspection of various garments. A fuzzy-tuned PID (FT-PID) control algorithm was employed to regulate the controller parameters automatically. The simulation and experimental results showed that FT-PID controller outperforms conventional PID controller and that it could provide efficient performance even when the force sensor output is contaminated with noise.

3.3 Assessing Seam Pucker

Seam pucker is still a primary concern in garment manufacturing. When the sewing parameters and material properties are not properly selected, puckering appears like a wave front along the seam line of a garment and damages its aesthetic value. This problem occurs immediately after seam construction or may develop after several washing and drying processes (Ukponmwan et al. 2000).

For quality garments, it is important to accurately evaluate seam pucker to better understand its causes and to ultimately eliminate it in the garment manufacturing process. The initial methods of seam pucker evaluation were based on subjective assessment; they suffered from the limitations of higher evaluation time, inconsistency among judges, and need for training; the results are not reliable. Therefore, different approaches to assess seam puckering objectively have been done in order to accurately rate the level of puckering in the sewn fabric. Among these approaches, artificial neural network, neuro-fuzzy logic, and machine learning methods look promising. Park and Kang (1997) presented an objective method for evaluating seam pucker in woven fabrics during garment manufacturing by using artificial neural networks. They showed that the neural networks evaluate seam pucker the same way as the AATCC standard rating of well-trained human experts. The authors highlighted that this method can be used to find good sewing parameters and suitable auxiliary materials, including sewing thread, interlining. In another studies, Park and Kang (1999a, b) developed a new quantitative method to evaluate seam pucker with five shape parameters using three-dimensional image analysis and neuro-fuzzy logic. The shape parameters included the number of wave generating points, the wave amplitudes, and the wavelengths on the line next to the seam and on the edge line. These parameters were not directly obtained from a simple analysis of the power spectra, but from fuzzy logic and neural networks in order to analyze power spectra in more detail and to produce shape parameters with the neuro-fuzzy algorithm. The authors pointed out that the new grading system can contribute to improved garment quality by identifying and solving seam pucker problems related to material properties as well sewing conditions. This method can be instrumental in better understanding the cause of seam pucker and eventually eliminating it in garment manufacturing plants (Park and Kang 1999c). Pavlinic et al. (2006) investigated the effect of fabric mechanical properties on the quality of seam appearance using machine learning methods including regression tree and k-nearest neighbors (K-NN). The obtained results indicated that the K-NN algorithm is more appropriate for the purpose than regression trees and that there is a high degree of correlation between the ranks of the attributes selected by the experts and those selected by the K-NN method. In addition, it was confirmed that fabric elasticity had the most prominent impact on seam puckering. In fact, it was found that seam puckering was more noticeable with inelastic fabrics, where the tension in warp and weft threads was higher than with elastic fabrics, due to the fact that they were pushed aside as each sewing needle penetration. The authors underlined that the proposed approach is of considerable importance in the process of designing high-quality garments and that it offers, apart from savings in the amount of the fabric to be used, clear criteria for the required fabric quality parameters.

Thus, the use of artificial intelligence techniques will be helpful for the manufacturers and customers to exactly evaluate the seam pucker and accordingly control the quality of apparels.

3.4 Detecting and Classifying Garments Defects

The main purpose of garment quality control is to identify the faults at the earliest possible steps for production of garments, and earlier the defects will be detected lesser will be the wastage of fabric, time, and money. However, identifying garment defects through visual inspection is not reliable. Therefore, automated defect detection and classification is required to enhance the product quality in order to meet both customer demands and to reduce the costs associated with off-quality. Hence, numerous approaches were proposed to address the problem of detecting defects in the fabric or garment (Mahajan et al. 2009; Ngan et al. 2011).

Neural networks are one of the fastest most flexible classifier used for fault detection due to their nonparametric nature and ability to describe complex decision regions. Bahlmann et al. (1999) proposed a NN-based system for an automated, vision-based quality control of textile seams with the aim to establish a standardized quality measure and to lower costs in manufacturing. The system could evaluate seam quality from grayvalue images. It consisted of a suitable image acquisition setup, an algorithm for locating the seam, a feature extraction stage, and a neural network of the self-organizing map type for feature classification. The obtained results showed that even with few but well-fashioned features good classification results could be obtained. The authors highlighted that the proposed system would be useful not just to objectify quality control of textile articles, but it can also provide a basis to perform online adjustment of sewing machine parameters to achieve smoother seams. Wong et al. (2009) presented a stitching and classification technique based on wavelet transform and BP neural network. Five classes of common stitching defect samples including pleats, puckers, tension, skipped-stitches, and hole were analyzed. The classification results demonstrated that the proposed method had high recognition accuracy in the detection and classification of stitching defects and that it exhibited better performance compared to wavelet-based methods. Yuen et al. (2009a) proposed a fabric stitching inspection method for knitted fabrics in which a segmented window technique was developed to segment images into three classes using a monochrome single-loop ribwork of knitted garment: (1) seams without sewing defects; (2) seams with pleated defects; and (3) seams with puckering defects caused by stitching faults. Nine characteristic variables were obtained from the segmented images and input into a BP neural network for classification and object recognition. The classification results demonstrated that the developed inspection method achieved high recognition accuracy and that it can provide decision support in defect classification (Yeun et al. 2009b). Also, it was proven that the classifier with nine characteristic variables outperformed those with five and seven variables and that the neural network technique using either BP or radial basis (RB) is effective for classifying the fabric stitching defects.

Kulkarni and Patil (2012) designed an automated garment identification and defect detection model based on texture feature, Gray-Level Co-occurrence Metrics (GLCM) and Probabilistic Neural Network (PNN). The texture features were used to detect garment defects, and these defects are classified by using PNN classifier. The experiment results showed that the proposed model is effective and suitable for online garment inspection and that PNN exhibited better detection accuracy compared with BP neural network.

In an automation inspection system for defect detection and classification, it is necessary to solve the problem of detecting small defects that locally break the homogeneity of a texture pattern and to classify different kinds of defects, including color value defects of fabrics. Zhang et al. (2011) developed a new intelligent and automated inspection model based on genetic algorithms and a modified Elman method neural network to detect and classify colored texture fabric defects that are also suitable for garment-stitching defects. The results demonstrated that the proposed inspecting model is more feasible and applicable in fabric defect detection and classification.

The main advantage of using an automated visual inspection system is that it does not suffer from limitations of humans, such as exhaustion, while offering the potential for robust defect detection, leading to reduced cost and time-wasting.

3.5 Dimensional Change Issue

Dimensional stability is regarded as being of primary importance to finished garments. A fabric or garment may exhibit shrinkage in some dimensions due to some relaxation process which enables the strains and distortions imposed on the fabric during manufacturing and processing to be released, allowing the fabric to take up a stable relaxed configuration (Kaulkanci and Kurumar 2015). Shrinkage is a combined effect of number of factors such as relaxation, finishing, dyeing, and effects of machinery (Kaur and Roy 2016). Garments made from fabrics without dimensional stability may change shape after laundering or dry-cleaning, which is undesirable for wearers. Kalkanci et al. (2017) investigate the use of a feed-forward back-propagation NN to estimate dimensional measure properties of T-shirts made up of single jersey and interlock fabrics. To that end, for each of the two fabric groups made up of different materials with three different densities either containing elastane or not, a total of 72 different types of T-shirts were manufactured. Knitted fabrics were processed through finishing operations in a garment controlled environment. Following the garment manufacturing process, dimensions of the ultimate product (T-shirt) were taken and recorded individually. The experimental results showed that the prediction of dimensional properties produced by the NN model was highly reliable ($R2 > 0.99$). The authors concluded that ANNs could successfully estimate dimensional changes in a garment and that they would eliminate additional operations to resolve the tension problem, thereby increasing productivity.

4 Garment Quality Evaluation

The garment shaped and manufactured so that it fits the 3D shape of the human body should meet the criteria of appearance quality and comfort in wearing. Indeed, consumer satisfaction with apparel products is influenced by physical as well as the psychological qualities of product. Hence, assessment of apparel product not only is limited to the functional aspects, but also includes the aesthetics.

In this section, we will present some applications of soft computing methods for evaluating clothing quality.

4.1 Clothing Sensory Comfort

Sensorial comfort is the sensation of how a fabric or garment feels when it is worn next to the skin. This feeling may be pleasant like smoothness or softness, or be unpleasant, if a textile is scratchy, too stiff, or clinging to a sweat-wetted skin (Nawaz et al. 2011). The most commonly used objective evaluation of this aspect of comfort is carried out by measuring the mechanical properties of apparel fabrics using the Kawabata Evaluation System (KES). However, the relationship between these properties and sensory data is so complex that traditional statistics cannot give accurate results. Thus, AI techniques such as NN and fuzzy logic can provide an attractive alternative to predict clothing sensory comfort. Wong et al. (2003) developed feed-forward BP neural network models to predict an overall comfort perception from ten individual sensory perceptions (clammy, clingy, damp, sticky, heavy, prickly, scratchy, fit, breathable, and thermal), which were rated by twenty-two professional athletes in a controlled laboratory. The obtained results showed a good agreement between predicted and actual clothing ratings. Also, it was shown that NN provided quick and flexible solutions with self-learning ability for such simulations compared with statistical modeling techniques.

Wong and Li (2004) investigated the process of human psychological perceptions of clothing-related sensations and comfort to develop an intellectual understanding of and methodology for predicting clothing comfort performance from fabric physical properties using different hybrid models. A series of running wear trial, which involved 8 sets of tight-fit garment and 28 subjects, was conducted in an environmentally controlled chamber. Thirty-three fabric physical property indexes were measured, and nine individual sensations (clammy, sticky, breathable, damp, heavy, prickly, scratchy, tight, and cool) and overall comfort were rated by the subjects during the running period. In the prediction of overall comfort ratings, fuzzy logic, linear model, and NN were employed separately in the different hybrid models. The obtained results showed that the model that integrates the three modeling techniques could generate the best predictions compared with other hybrid models. This finding was attributed to the fact that this model combines the strengths of NN (self-learning capability), fuzzy logic (fuzzy reasoning ability), and statistics (data reduction and information summation). The authors highlighted that fabric physical properties can be used to predict overall clothing comfort with the application of the hybrid predictive model that it can closely simulate human sensory perception and judgement processes. Wang et al. (2008) developed an expert system based on fuzzy logic to describe sensory on clothing in accordance with professional knowledge and consumer preference and showed that it could be applied for product designing and fashion trends tracing in garment industry.

4.2 Clothing Thermal Properties

The thermal comfort is related to the ability of fabric to maintain the temperature of skin through transfer of heat and perspiration generated within the human body. Nowadays, various consumers consider thermal comfort as one of the most significant attributes when purchasing apparel products. The thermal properties of fabrics have been objectively and subjectively evaluated by several techniques (Pamuk 2008). However, owing to nonlinear relationship of different fibers, yarn and fabric parameters with thermal properties, difficulty arises in the statistical modeling. Therefore, it becomes difficult to study the effect of some parameters without varying the other parameters. Thereby, advanced techniques such as fuzzy logic and ANN look promising. Wang et al. (2005) presented a fuzzy system to predict the subjective perceptions of thermal comfort on the basis of simulated results of thermal and moisture sensations, in which dampness and coolness sensations were considered to be two main factors that would affect the final, clothing thermal comfort perception. The membership functions of these, as well as inference rules, were established through data collected from questionnaires in a series of wear trials carried out in a climate chamber. The experiments were then simulated again by specifying experimental conditions, human physiological activity level, as well as the textile material used. During the simulation, the following information was calculated: dynamic temperature and moisture concentration distribution of the human-clothing-environment system, the firing rate of the thermal receptors, the perceptions of dampness and coolness due to the contact of clothing, and the overall perception of clothing thermal comfort. It was found that the simulation results agreed well with the experimental results. Luo et al. (2007) developed a fuzzy NN thermal comfort model for evaluating the apparel thermal function in a dynamical and non-uniform environment, which was determined by the wearer's local and overall thermal sensations using physiological parameters including core and local body part temperatures and the rates of temperature changes. The test results for simulation data verified the reliability of this human-like approach. Al-Rashidi et al. (2015) used an ANN to predict the thermal insulation values of children's school wear in Kuwait classrooms. The obtained results showed that ANN was able to give more accurate prediction of the clothing thermal insulation values than regression equation and standard tables methods. The weight of each variable in the neural network structure was used to estimate the relative importance of each variable on the clothing thermal insulation prediction, and results indicated that the weight of the cloths had the most pronounced effect on the thermal insulation value. The authors indicated that the findings of their study give evidence about the applicability of the new ANN model to predict the thermal insulation of different children's clothing ensembles.

4.3   Garment Appearance Quality

Garment appearance is an important factor in determining clothes aesthetic aspect. Fabric's mechanical properties have prominent influence on fabric behavior during garment manufacturing and final product shaping (Mousazadegan et al. 2013). Formability of the fabrics in particular garment manufacturing processes and the stability of the newly created form directly impact garment appearance quality. Furthermore, an important role in ensuring the quality of the garment made is played by the additional factors, such as drape. These factors cannot be measured, but can be evaluated employing subjective grades, visual by nature. However, the subjective method does not offer engineering assessment of garment quality. Thus, with increasing market requirements, it is necessary to develop objective methods to evaluate fabric appearance quality. Pavlinic and Gersak (2009) presented an intelligent system for predicting garment appearance quality, based on studying the interactions between the parameters of fabric mechanical and physical properties, as measurable values, and the grade of garment appearance quality, i.e., the grade of each individual garment appearance factor, expressed by descriptive subjective grades. The factors of garment appearance quality were determined on the basis of the elements for obtaining proper drape, i.e., garment yield, achieving 3D shape, garment fit, the quality of the seams made, and the quality of garment appearance as a whole. A group of semiskilled evaluators and a group of experts in the field of garment engineering were doing the evaluation. The method of nearest neighbor k-NN was used to design the prediction model, showing better accuracy than regression trees. The authors highlighted that the developed intelligent system is of a particular applicative importance, as it can be used in engineering predictions and designing high-quality garments, while at the same time it offers important data on the quality requirements of particular parameters of fabric mechanical and physical properties, necessary to obtain the required garment appearance. They also outlined that this system represents a necessary objective technology of measuring and evaluating garment appearance quality, since the existing conventional methods of subjectively evaluating garment appearance quality should be replaced by a new knowledge-based engineering method.

Over the years, predicting fabric behavior during garment manufacturing process was considered by researchers in order to reduce manufacturing problems and achieve high-quality products. Xue et al. (2016) developed an ANFIS predictive model to study the relations between fabric formability and the mechanical properties for the end-use of men's suits. Based on the integration of NN and fuzzy inference, the proposed model has proven to be capable of producing results of higher predictive precision and better interpretability as compared with classical methods. The authors highlighted that the results obtained from their research are believed to be valuable for suit manufacturers and researchers who are working on the translation between fabric physical properties and the desired silhouette of men's suits and, meanwhile, could be instructive for dealing with many other similar problems concerning data uncertainty and imprecision.

Drape simulation is a very challenging task. Fan et al. (2001) investigate the use of a fuzzy-neural network system to predict and display the drape image of garments made from different fabrics and styles. The basic logic was to find and display a drape image from a database that was very close to the actual drape image of the newly designed garment of the same style. The proposed system has been tested to be satisfactory with lady's dress made of a wide range of fabrics. The advantages of the approach included very fast computation, avoiding difficulty of taking into account the effects of accessories, seams, and styles on drape in conventional drape simulation, and if sufficient drape images are stored in the database, the predicted drape image can be very close to the actual one. A disadvantage was that only limited styles and changeable feature dimensions can be allowed in the approach.

5 Challenges Facing Adoption of AI Techniques in Clothing Industry

Published literature presented in this chapter show the potential of AI techniques for providing support in decision-making and problem-solving processes involved in garment manufacturing. Nevertheless, in spite of these advantages, clothing companies do not widely use these advanced techniques. This fact can be attributed to the followings challenges and limitations:

- The lack of data availability and limited sample sizes. In fact, current AI approaches require a lot of labeled data in order to achieve decent accuracy in their predictions. However, apparel enterprises lack sufficient training data that are labeled, since labeling often requires expensive human labor and much time, which means the solutions fall short. Thus, AI techniques need to evolve toward *Unsupervised Learning* models that do not require labeled data to train the AI models.
- The cost of incorporating artificial intelligence in daily operations is still very high, and further advancements in the technology might bring that down. However, the investment and resources required to produce intelligent machines that can perform complex human tasks may not be justified by the increased productivity or cost savings.
- Most of the previous studies focused on presenting an AI-based methodology to handle a specified problem, and only few papers have compared the performance of different AI techniques.
- The long computational time required for handling a large size of data set. Indeed, easy or traceable problems can be solved in polynomial time. However, intractable problems require times that are exponential functions of the problem size.
- The setting of parameters of an AI technique has a large effect on its performance. Due to the insufficiency of theoretical foundation of AI techniques, parameter setting usually depends on experience or the trial-and-error method in previous studies.

- The effects of changes in many processing variables in the manufacturing process of garments have not been fully quantified. This is an important task for the clothing industry to accomplish in the future.
- The decision-making problems in the apparel industry have many distinct features, such as more uncertainties and dynamic features, which increase the complexity of these problems. In previous studies, various practical features existing in the clothing industry such as machine breakdown, variable operator efficiencies, and operator absenteeism were not investigated.
- The lack of confidence in the fairness of an AI-based system will limit support for its use and likely preclude adoption, even if that adoption could provide significant benefits.
- The lack of insight into how these systems work in the first place. For example, neural networks are usually inscrutable to observers and act as a black box, which does not consider nor explain the underlying physical processes explicitly. Although we know how they are put together and the information that goes in them, the reasons why they come to certain decisions usually goes unexplained.
- Today's expert systems have no ability to learn from their experience. Except for simple classification systems, expert systems do not employ a learning component to construct parts of their knowledge bases from libraries of previously solved cases. However, learning capabilities are needed for intelligent systems that can remain useful in the face of changing environments or changing standards of expertise.
- AI interference in human roles can cost jobs for a considerable share of our manpower. Hence, critics might lobby against extended AI implementation.
- Cultural resistance to change also high on the list of practical challenges provided by respondents will be a tougher nut to crack for many apparel companies.

6 Conclusion

AI offers great potential for the engineering of garments since it is able to significantly reduce the product development process and lead to great savings. It becomes an excellent tool for decision makers to effectively plan and execute complex manufacturing tasks, rapidly identify and solve critical quality issues, and deliver products that satisfy unique customer requirements.

In this chapter, the research on applications of AI in garment manufacturing was examined and classified into three categories as production planning, control, and scheduling; garment quality control and inspection; and garment quality evaluation. The findings reveal that artificial neural networks, genetic algorithms, fuzzy logic, neuro-fuzzy systems, and machine learning methods are the most frequent AI tools used in the applications in garment manufacturing. Another noteworthy finding is that the research issues investigated in previous studies are limited: A number of problems have rarely been discussed or examined.

Furthermore, the industrial use of AI technologies in the clothing industry is still very limited. This is because important hurdles exist at various levels. Thus, the implementation of the AI techniques into the clothing industry requires a careful consideration of the various practical features existing in the clothing industry in order to ensure optimal solutions. In addition, the effects of various AI model parameters, such as parameters of ANN structures, on the performance of solutions should be investigated in order to provide definite methods to set these parameters for certain problems. Meanwhile, AI with all its challenges and opportunities is an inevitable part of our future. Therefore, both the AI researchers and the apparel industry experts need to work closely together in order to explore practical decision-making problems not yet investigated in the apparel industry and improve the efficiency of existing AI approaches.

References

Abeysooriya RP, Fernando TGI (2012a) Canonical genetic algorithm to optimize cut order plan solutions in apparel manufacturing. J Emerg Trends Comput Inf Sci 3(2):150–154

Abeysooriya RP, Fernando TGI (2012b) Hybrid approach to optimize cut order plan solutions in apparel manufacturing. Int J Inf Commun Technol Res 2(4):348–353

Adeli H (2003) Expert systems in construction and structural engineering. CRC Press. ISBN 9780203401101

Al-Rashidi K, Alazmi R, Alazmi M (2015) Artificial neural network estimation of thermal insulation value of children's school wear in Kuwait classroom. Adv Artif Neural Syst Article ID 421215, 9 pp. http://dx.doi.org/10.1155/2015/421215

Bahlmann C, Heidemann G, Ritter H (1999) Artificial neural networks for automated quality control of textile seams. Patt Recogn 32(6):1049–1060

Barrett GR, Clapp TG, Titus KJ (1996) An on-line fabric classification technique using a wavelet-based neural network approach. Text Res J 66(8):521–528

Bouziri A, M'hallah R (2007) A hybrid genetic algorithm for the cut order planning problem. In: New trends in applied artificial intelligence. Lecture notes in computer science, vol 4570, pp 454–463

Brown EC, Sumichrast RT (2005) Evaluating performance advantages of grouping genetic algorithms. Eng Appl Artif Intell 18:1–12

Carvalho H, Silva LF, Soares F, Guhr F (2010) Adaptive control of an electromagnetically presser-foot for industrial sewing. In: 2010 IEEE 15th conference on emerging technologies and factory automation (ETFA 2010), 13–16 September, Bilbao, Spain

Chan CC, Hui CL, Yeung KW, Ng SF (1998) Handling the assembly line balancing problem in the clothing industry using a genetic algorithm. Int J Cloth Sci Technol 10(1):21–37

Chen R-S, Lu K-Y, Yu S-C (2002) A hybrid genetic algorithm approach on multi-objective of assembly planning problem. Eng Appl Artif Intell 15(5):447–457

Chen JC, Hsiao MH, Chen CC, Sun CJ (2009) A grouping genetic algorithm for the assembly line balancing problem of sewing lines in garment industry. In: 2009 international conference on machine learning and cybernetics, Hebei, China

Chen CJ, Chen C-C, Su L-H, Sun C-J (2012) Assembly line balancing in garment industry. Expert Syst Appl 39(11):10073–10081

Chen JC, Chen CC, Lin YJ, Lin CJ, Chen TY (2014) Assembly line balancing problem of sewing lines in garment industry. In: Proceedings of the 2014 international conference on industrial engineering and operations management, Bali, Indonesia, 7–9 January 2014, pp 1215–1225

Cooklin G, Hayes SG, McLoughlin J (2006) Introduction to clothing manufacture. Blackwell Publishing Ltd., pp 85–99

Du W, Tang Y, Yung S et al (2017) Robust order scheduling in the fashion industry: a multi-objective optimization approach. CoRR abs/1702.00159

Dumishllari E, Guxho G (2016) Influence of lay plan solution in fabric efficiency and consume in cutting section. AUTEX Res J 16(4):222–227

Eryuruk SH, Kalaoglu F, Baskak M (2008) Assembly line balancing in a clothing company. Fibres Text East Eur 16(1):93–98

Falkenauer E (1993) The grouping genetic algorithm: widening the scope of the Gas. JORBEL Belg. J Oper Res Stat Comput Sci 33:79–102

Fan J, Newton E, Au R, Chan SCF (2001) Predicting garment drape with a fuzzy-neural network. Text Res J 71(7):605–608

Fung EHK, Wong YK, Zhang XZ, Cheng L, Yuen CWM, Wong WK (2011) Fuzzy logic control of a novel robotic hanger for garment inspection: modeling, simulation and experimental implementation. Expert Syst Appl 38:9924–9938

Gong RH, Chen Y (1999) Predicting the performance of fabrics in garment manufacturing with artificial neural networks. Text Res J 69(7):477–482

Guhr F, Silva L, Soares F and Carvalho H (2004) Fuzzy logic based control strategies for an electro-magnetic actuated sewing machine presser foot. In: Proceedings of the 2004 IEEE international conference on industrial technology, pp 985–990

Guo ZX, Wong WK, Leung SYS, Fan JT, Chan SF (2006) Mathematical model and genetic optimization for the job shop scheduling problem in a mixed- and multiproduct assembly environment: a case study based on the apparel industry. Comput Ind Eng 50:202–219

Guo ZX, Wong WK, Leung SYS, Li M (2011) Application of artificial Intelligence in the apparel industry: a review. Text Res J 5(12):1871–1889

Guo Z, Wong WK, Leung S (2013) A hybrid intelligent model for order allocation planning in make-to-order manufacturing. Appl Soft Comput 13(3):1376–1390

Hopper E, Turton BCH (2001) An empirical investigation of meta-heuristic and heuristic algorithms for a 2D packing problem. Eur J Oper Res 128(1):34–57

Huang G (2013) Application of optimized particle swarm algorithm on apparel intelligent layout. Appl Mech Mater 380–384:1668–1672

Hui CL, Ng SF (2005) A new approach for prediction of sewing performance of fabrics in apparel manufacturing using artificial neural networks. J Text Instit 96(6):401–405

Hui CL, Ng SF (2009) Predicting seam performance of commercial woven fabrics using multiple logarithm regression and artificial neural networks. Text Res J 79(18):1649–1657

Islam MM, Mohiuddin HM, Mehidi SH, Sakib N (2014) An optimal layout design in an apparel industry by appropriate line balancing: a case study. Glob J Res Eng G Ind Eng 14(5):35–43

Kalkanci M, Kurumer G, Ozturk H, Sinecen M, Kayacan O (2017) Artificial neural network system for prediction of dimensional properties of cloth in garment manufacturing: case study on a T-shirt. Fibres Text East Eur 25(4):135–140

Kaulkanci M, Kurumar G (2015) Investigation of dimensional changes during garment production and suggestions for solutions. Fibers Text East Eur 23(3):8–13

Kaur A, Roy K (2016) Prediction of shrinkage and fabric weight (g/m^2) of cotton single jersey knitted fabric using artificial neural network and comparison with general linear model. Int J Inf Res Rev 2541–2544

Kim I, Fok S, Fregene K, Lee D, Oh T, Wang D (2004) Neural network-based system identification and controller synthesis for an industrial sewing machine. Int J Control Autom 2:83–91

Koustoumpardis P, Aspragathos N (2003) Fuzzy logic decision mechanism combined with a neuro-controller for fabric tension in robotized sewing process. J Intell Robot Syst 36:65–88

Koustoumpardis P, Aspragathos N (2007) Neural network force control for robotized handling of fabrics. In: Proceedings of the 2007 international conference on control, automation and systems, Seoul, South Korea. IEEE, pp 2845–2850

Kulkarni AH, Patil SB (2012) Automated garment identification and defect detection model based on texture features and PNN. http://www.ijltet.org/wp-content/uploads/2012/07/61. Accessed 29 Sept 2017

Lin MT (2009) The single-row machine layout problem in apparel manufacturing by hierarchical order-based genetic algorithm. Int J Cloth Sci Tech 20(5):258–270

Luo X, Hou W, Li Y, Wang Z (2007) A fuzzy neural network model for predicting clothing thermal comfort. Comput Math Appl 53:1840–1846

M'hallah R, Bouziri A (2016) Heuristics for the combined cut order planning two-dimensional layout problem in the apparel industry. Int Trans Oper Res 23:321–353

Mahajan PM, Kolhe SR, Patil PM (2009) A review of automatic fabric defect detection techniques. Adv Comput Res 1(2):18–29

Martens J (2004) Two genetic algorithms to solve a layout problem in the fashion industry. Eur J Oper Res 154(1):304–322

Mok PY, Kwong CK, Wong WK (2007) Optimization of fault-tolerant fabric cutting schedules using genetic algorithms and fuzzy set theory. Eur J Oper Res 177:1876–1893

Mok PY, Cheung TY, Wong WK et al (2013) Intelligent production planning for complex garment manufacturing. 24(1):133–145

Mousazadegan F, Ezazshahabi N, Latifi M, Saharkhiz S (2013) Formability analysis of worsted woven fabrics considering fabric direction. Fibers Polym 14(11):1933–1942

Naik SB, Kallurkar S (2016) A literature review on efficient plant layout design. Int J Ind Eng Res Dev (IJIERD) 7(2):43–51

Nascimento DB, Figueiredo JN, Mayerle SF, Nascimento PR, Casali RM (2010) A state-space solution search method for apparel industry spreading. Int J Prod Econ 128(1):379–392

Nawaz N, Troynikov O, Watson C (2011) Evaluation of surface characteristics of fabrics suitable for skin layer of firefighters' protective clothing. Phys Procedia 22:478–486

Ngan HYT, Pang GKH, Yung NHC (2011) Automated fabric defect detection—a review. Image Vis Comput 29(7):442–458

Onal L, Zeydan M, Korkmaz M, Meeran S (2009) Predicting the seam strength of notched webbings for parachute assemblies using the Taguchi's design of experiment and artificial neural networks. Text Res J 79(5):468–478

Ozel Y, Kayar H (2008) An application of neural network solution in the apparel industry for cutting time forecasting. In: 8th WSEAS international conference on simulation, modelling and optimization (SMO '08), Santander, Cantabria, Spain, 23–25 September 2008, pp 224–218

Pamuk O (2008) Clothing comfort propeties in textile industry. e-J New World Sci Acad 3(1):69–74

Park CK, Kang TJ (1997) Objective rating of seam pucker using neural networks. Text Res J 67(7):494–502

Park CK, Kang TJ (1999a) Objective evaluation of seam pucker using artificial intelligence, Part I: Geometric modeling of seam pucker. Text Res J 69(10):735–742

Park CK, Kang TJ (1999b) Objective evaluation of seam pucker using artificial intelligence, Part II: Method of evaluating seam pucker. Text Res J 69(11):835–845

Park CK, Kang TJ (1999c) Objective evaluation of seam pucker using artificial intelligence, Part III: Using the objective evaluation method to analyze the effects of sewing parameters on seam pucker. Text Res J 69(12):919–924

Patrick CLH, Frency SFN, Keith CCC (2000) A study of the roll planning of fabric spreading using genetic algorithms. Int J Cloth Sci Technol 12(1):50–62

Patrick CLH, Keith CCC, Yeung KW, Frency SFN (2007) Application of artificial neural networks to the prediction of sewing performance of fabrics. Int J Cloth Sci Technol 19(5):291–318

Pavlinic DZ, Gersak J (2009) Predicting garment appearance quality. Open Text J 2:29–38

Pavlinic DZ, Gersak J, Demsar J, Bratko I (2006) Predicting seam appearance quality. Text Res J 76(3):235–242

Rose DM, Shier DR (2007) Cut scheduling in the apparel industry. Comput Oper Res 34(11):3209–3228

Silva LF, Carvalho H, Soares F (2004) Improving feeding efficiency of a sewing machine by on-line control of the presser-foot. In: Proceedings of the 4th international conference on advanced engineering design—AED'2004 (CD-ROM), Glasgow, Scotland, UK, 5–8 September 2004

Ukponmwan JO, Mukhopadhyay A et al (2000) Sewing thread. Text Inst (Manchester) 30:1–91

Ultutas B, Islier AA (2015) Dynamic facility layout problem in footwear industry. J Manuf Syst 36:55–61

Unal C, Tunali S, Guner M (2009) Evaluation of alternative line configurations in apparel industry using simulation. Text Res J 79(10):908–916

Vorasitchai S, Madarasmi S (2003) Improvements on layout of garment patterns for efficient fabric consumption, circuits and systems. In: ISCAS '03, Proceedings of the 2003 IEEE international symposium on circuits and systems, 25–28 May 2003, Bangkok, Thailand

Wang Z, Li Y, Wong A (2005) Simulation of clothing thermal comfort with fuzzy logic. Elsevier Ergon Book Ser 3:467–471

Wang L, Chen Y and Wang Y (2008) Formalization of fashion sensory data based on fuzzy set theory. In: Guo M, Zhao L and Wang L (eds) Proceedings of the 4th international conference on natural computation, Jinan, China. IEEE Computer Society, pp 80–84

Wong WK (2003) Optimisation of apparel manufacturing resource allocation using a generic opti-mised table-planning model. Int J Adv Manuf Technol 21(12):935–944

Wong WK, Chan CK (2001) An artificial intelligence method for planning the clothing manufac-turing process. J Text Inst 92(2):168–178

Wong WK, Leung SYS (2008) Genetic optimization of fabric utilization in apparel manufacturing. Int J Prod Econ 114(1):376–387

Wong WK, Leung SYS (2009) A hybrid planning process for improving fabric utilization. Text Res J 79(18):1680–1695

Wong ASW, Li Y (2004) Prediction of clothing comfort perceptions using artificial intelligence hybrid models. Text Res J 74(1):13–19

Wong ASW, Li Y, Yeung PKW, Lee PWH (2003) Neural network predictions of human psycho-logical perceptions of clothing sensory comfort. Text Res J 73(1):31–37

Wong WK, Kwong CK, Mok PY, Ip WH, Chan CK (2005a) Optimization of manual fabric-cutting process in apparel manufacture using genetic algorithms. Int J Adv Manuf Technol 27(1–2):152–158

Wong WK, Leung SYS, Au KF (2005b) A real-time GA-based rescheduling approach for the pre-sewing stage of an apparel manufacturing process. Int J Adv Manuf Technol 25(1–2):180–188

Wong WK, Kwong CK, Mok PY, Ip WH (2006a) Genetic optimization of JIT operation schedules for fabric-cutting process in apparel manufacture. J Intell Manuf 17:341–354

Wong WK, Mok PY, Leung SYS (2006b) Developing a genetic optimisation approach to balance an apparel assembly line. Int J Adv Manuf Technol 28(3/4):387–394

Wong WK, Yuen CWM, Fan DD, Chan LK, Fung EHK (2009) Stitching defect detection and classification using wavelet transform and BP neural network. Expert Syst Appl 36:3845–3856

Wong WK, Wang XX, Guo ZX (2013a) Optimizing marker planning in apparel production using evolutionary strategies and neural networks. In: Optimizing decision making in the apparel supply chain using artificial intelligence (AI): form production to retail. Woodhead Publishing Series in Textiles, pp 106–131

Wong WK, Mok PY, Leung SYS (2013b) Optimizing apparel production systems using genetic algorithms. In: Optimizing decision making in the apparel supply chain using artificial intelligence (AI): form production to retail. Woodhead Publishing Series in Textiles, pp 153–169

Wong WK, Guo Z, Leung S (2014) Intelligent multi-objective decision-making model with RFID technology for production planning. Int J Product Econ 147(Part C):647–658

Xue X, Zeng X, Koehl L (2016) An intelligent method for the evaluation and prediction of fabric formability for men's suits. https://doi.org/10.1177/0040517516681956

Yeun CWM, Wong WK, Qian SQ, Fan DD, Chan LK, Fung EHK (2009a) Fabric stitching inspection using segmented window technique and BP neural network. Text Res J 79(1):24–35

Yeun CWM, Wong WK, Qian SQ, Chan LK, Fung EHK (2009b) A hybrid model using genetic algorithm and neural network for classifying garment defects. Expert Syst Appl 36(2):2037–2047

Yildiz Z, Dal V, Ünal M, Yildiz K (2013) Use of artificial neural networks for modelling of seam strength and elongation at break. Fibres Text East Eur 21(5):117–123

Yolmeh A, Kianfar A (2012) An efficient hybrid genetic algorithm to solve assembly line balancing problem with sequence dependent setup times. Comput Ind Eng 62(4):936–945

Zacharia P (2012) Robot handling fabrics towards sewing using computational intelligence methods. In: Dutta A (ed) Robotic systems—applications, control and programming. ISBN 978-953-307-941-7. https://doi.org/10.5772/25918

Zacharia P, Aspragathos NA, Mariolis I, Dermatas E (2009) Robotic system based on fuzzy visual servoing for handling flexible sheets lying on a table. Ind Robot Int J 36(5):489–496

Zhang YH, Yuen CWM, Wong WK, Chi-wai Kan (2011) An intelligent model for detecting and classifying color-textured fabric defects using genetic algorithms and the Elman neural network. Text Res J 81(17):1772–1787

AI for Apparel Manufacturing in Big Data Era: A Focus on Cutting and Sewing

Yanni Xu, Sébastien Thomassey and Xianyi Zeng

Abstract In the fashion industry, the apparel manufacturing part contains four main processes involving cutting, sewing, finishing, and packing. The complex system deals with configuration of numerous operations and resources in facing of various uncertainties and under constraints of sequence, quantity, time, and cost. Artificial intelligence (AI) has been applied to provide optimal scenario in shorter time than traditional mathematical methods. Big data is helpful due to the ability of prediction for unraveling uncertainties which ensures a smooth and stable production. For improvement of apparel manufacturing in modern fashion industry, it is necessary to develop the capabilities of advanced computing technologies and take great advantage of valuable information that can be dug out from big data.

Keywords Artificial intelligence · Big data · Apparel manufacturing · Review

1 Introduction

Apparel manufacturing is a subsection of fashion industry, as shown in Fig. 1.

Fashion industry plays an important role in daily life since it produces what people wear. Likewise, it has great contribution to global economics since the whole industry is very large that it is related to the international clothing supply chain from raw materials to ready-to-wear including two main sections: making and selling. In the making part, an apparel should first be designed and then be put into manufacturing where the manufacturing section consists of two subsections: raw material manufacturing and apparel manufacturing. It is the apparel manufacturing process that turns the fabric into the final garments. The traditional apparel manufacturing is characterized by labor-intensive and low technology where lays an imperative requirement of advanced scientific technologies to replace those outdated methods. Well progress has been seen in certain areas like computer-aided production

Y. Xu (✉) · S. Thomassey · X. Zeng
ENSAIT-GEMTEX, 2 allée Louise et Victor Champier, 59100 Roubaix, France
e-mail: yanni.xu@ensait.fr

© Springer Nature Singapore Pte Ltd. 2018
S. Thomassey and X. Zeng (eds.), *Artificial Intelligence for Fashion Industry in the Big Data Era*, Springer Series in Fashion Business,
https://doi.org/10.1007/978-981-13-0080-6_7

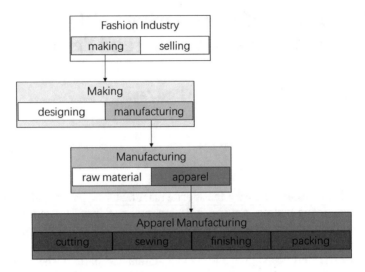

Fig. 1 Structure of fashion industry

planning, automated production process, and digitized logistics management. Especially in the past two decades, an increasing number of AI technologies are applied in modern fashion industry because of their high performance in solving the intractable decision-making problems where apparel manufacturing has attracted more attention from AI researchers than other fields in fashion industry (Guo et al. 2011). Developed science and technology promote industrial automation and computerization, which in turn leads to information growth that can increase the use of data-intensive methods. Consequently, it becomes much easier to better deal with real data collected from actual production by AI technologies. In this paper, perspectives of AI for apparel manufacturing in the big data era are proposed following a full awareness of the state of the art.

The remainder of the paper is organized as follows. Section 2 provides an introduction to the apparel manufacturing process through describing the main links involved and corresponding issues to each link. In addition, key problems and difficulties to cope with are illustrated and some generally used solutions are presented. Section 3 is a review of the work related to the use of AI-based methodologies in apparel manufacturing accompanied with analyses and conclusions. And finally, specific suggestions for future studies are proposed in the last section.

Significant contributions of the current review are: (1) Literature searching is not limited to the preliminary literature screening in database according to search criteria including publish year, keywords, and article type, but provides a broader coverage by tracking the related references and citations until there are no more suspects which are the guarantee of a searching result with accuracy and comprehensiveness. (2) A systematic analysis of the literature is provided through professional classification schemes updated and extended from the existing classification framework based on

Table 1 Classification scheme of issues in apparel manufacturing process

Section	Subsection	Issue	
Cutting	Production planning	Cutting order planning	Lay planning
			Marker making
		Spreading and cutting sequencing	
	Spreading	Roll sequencing	
		Fabric spreading	
	Cutting operation	Cutting route planning	
	Sorting and bundling	Sorting and bundling scheduling	
Sewing	Production planning	Sewing order planning	
		Line design	Layout design
			Sewing assembly line balancing
	Quality control	Products inspection	
		Defects prediction	
	Sewing operation	Fabric handling	
Finishing	Finishing operation	Fabric folding	
Packing	Packing operation	Products packaging	
Whole	Production planning	Order planning	
		Resource allocation	
	Quality control	Production monitoring	

actual production. (3) It offers a comprehensive summary of the present research status and points out the prospective tendency of the future study in detail.

2 Apparel Manufacturing Process

From fabric roll to ready-to-wear, apparel manufacturing process includes two main procedures: cutting and sewing. Cutting is the upstream process of sewing, while sewing is served by cutting. In terms of work division, cutting is to cut cut-pieces from fabric rolls, and sewing is to assemble cut-pieces into a ready-to-wear. The garments then will be delivered from the final workstation of the assembly line which is for inspection to the finishing section for ironing process and packaged into boxes in the end. Cutting, sewing, finishing/ironing, and packing are regarded as the four conventional sections in apparel manufacturing process. Table 1 illustrates a classification scheme of the issues in each section of apparel manufacturing process.

Production planning and quality control (QC) are quite essential subsections when considering the whole apparel manufacturing process (Guo et al. 2015; Lee et al. 2014a, 2016). The related issues can be solved similarly as for each section. For each link, they also have specific issues to deal with. Since the last two sections, finishing and packing, are in much smaller-scale and operations are not that complex compared with the other two sections, in real-life production they are located in one place and produce a unified whole. As a result, main links and the related issues are described here in three parts.

2.1 Cutting

Cutting is a series of work start with production planning which is constituted of cutting order planning (COP) and spreading and cutting sequencing (SCS), where the former consists of lay planning and marker optimization, then followed by spreading, cutting, and ends with sorting and bundling.

COP or called cutting scheduling is the basic and crucial subsection in cutting process, constituted of a set of fabric lays with relevant markers satisfying all constraints such as quantity and time demands of downstream sewing department, capacity of cutting equipment (spreading table and cutting machine), and labor force. A lay/stack consists of certain number of fabric plies, while a maker is the layout of patterns to be cut out from the lay. SCS refers to realize the spreading and cutting work balance within the spreading and cutting capacity. Spreading is the preparation before the cutting operation, which first is to spread fabric from rolls into a lay on cutting table and then locate the marker across the top fabric ply. Cut-pieces are cut out along the cutting route, from fabric lays during the cutting operation, and at last sorted and bundled for assembly use.

The cutting process should firstly fulfill the quantities of the cut-pieces required by the downstream sewing line on time. In addition, it is subject to physical constraints: the thickness of the cloth and the depth of the cutting knife which decide the maximum number of plies, the length of the cutting table which decides the maximum length of marker, widths of fabric rolls which decide the width of the marker. Abovementioned constraints make cutting planning, composed of COP and SCS, a complicated issue.

Multiple objectives involved in the cutting section are: reduction of the production costs regarding materials (fabric and marker), higher utilization of resources (labor and machine), greater accuracy, faster throughput, minimal excessive production, and lowest inventory.

Solutions for satisfying the constraints and achieving the objectives are listed below:

(1) Apart from the fabric loss due to fabric flaws, there are two causes of fabric loss in cutting process: One is the marking loss because of the unused areas on the markers, and the other is the spreading loss including end loss, width loss, splicing loss, and remnant loss (Hui et al. 2000). Increase of marker efficiency and proper spreading roll sequence are solutions to minimize fabric wastage.

(2) The work order needs to be covered by the shortest sequence of markers assuming the minimum consumption of materials for extra markers, and consequently,

the minimum time is spent on setup in spreading and cutting operation (Fister et al. 2010). A well-planned order of work should avoid earliness and lateness. The reason is that earliness causes extra inventory management cost, while lateness results in a delay of cut-piece input to the downstream sewing process.

(3) To figure out the shortest cutting path for cutting operation is helpful to reduce operation time and avoiding overtime work which contributes to machine and labor cost savings.

(4) Cutting and spreading operation balancing is for minimizing makespan/completion time for it avoids imbalance, the cause of idle time. Also, it brings the higher utilization of resources. Not only operations but also uncertainties in real dynamic cutting room should be considered in SCS, since uncertain factors such as late receipt of fabric rolls, insertion of rush orders, processing time deviations/variety of operator's skill level, machine breakdowns can lead to imbalance of work process (Wong 2003a).

(5) Robotic manufacturing is expected for improvement of efficiency and accuracy in cutting-related operations: spreading, cutting, as well as sorting and bundling. It could be considered costly at early stage, however, with the rapid increase of labor cost, in the long run, it must be a good deal.

In order to save fabric, advanced computing technologies could be applied in marker optimization, spreading roll management; also to save time and increase utilization of resources (labor and machine), they could be used in cutting order sequencing, spreading and cutting operations balancing, cutting path optimization, and machine intelligence development.

2.2 Sewing

Sewing is the most critical and intricacy section of apparel manufacturing process, since it deals with a large number of various operations, the relevant operators, and machines.

In the sewing process, work begins with production planning: One is sewing order planning, and the other is sewing assembly line design. Order planning of sewing process provides the production schedule for each sewing order to be put into production in sequence. The design of an assembly line, by contrast, covers more details as it includes two parts: layout design and sewing assembly line balancing (SALB). Assembly line is a sequence of workstations (equipped by operators who have the skills and technological capabilities required and machines with the required functions) connected by means of conveyance. Garment assembly lines would be in different shapes like straight line, serpentine/Z-shaped line, U-shaped line, or in a loop. Operators could face the same or opposite direction. Center tables or tools like trolley, basket, or hanger are used for material handling in different sewing systems, for example the Progressive Bundle System (PBS) and the Unit Production System (UPS) (Guo et al. 2006). These are elements considered in terms of the layout design. Line balancing is to distribute tasks evenly to workstations with machines equipped. Before production, the sewing line supervisors tackle the resources allocation problem with material, operator, and machine to achieve a balanced loading. In the sewing production line, each operator operates the given one or few fixed tasks in workstations equipped with sewing machines or ironing machines as materials moving across the workstations. For quality assurance, there will be inspections inserted in the middle and end of the sewing line.

Order planning of sewing process should take consideration of the task response from upstream cutting process and downstream finishing and packing process, and help to satisfy the due date of each order as well as decrease inventories and maximize resources utilizations. A good layout should be clever to reduce material transportation in order to save time and increase machine utilization. SALB is not easy work for it concerns a plenty variety of tasks, operators, and facilities. Exactly, SALB is regarded as the NP-hard problem. Variation of any link in a long chain would result in imbalance, especially in a long chain. For this reason, uncertainties are more troublesome to SALB. During sewing operation, there exist various uncertainties involving: the variety of operators' skill level, operator absenteeism, machine breakdowns, inserted rush orders. Sewing process is characterized by labor-intensive. Due to the complexity of sewing operations, the garment assembly work still relies mainly on manual operations subject to operators' skills. On aspect of quality control, a low rejection rate is desired as it contributes to the increase of production efficiency. Similarly, defective products are unwanted which lead to overwork caused by rework or repair.

Multiple objectives of sewing contain short lead time, high resources utilization (material, operator, and machine), low inventory, and high quality.

Solutions for better performance of sewing section are as follows:

(1) Orders should be produced in sequence within the limited time with least inventory. A well-arranged schedule of sewing orders could help.
(2) The assembly line layout should service material transportation that the path should be shortened to reduce no-value-added transmission and avoid crossing which will likely cause mistakes.
(3) Multi-skilled operators benefit flexibility of task assignment in assembly line balancing before production. To cope with unanticipated interference in the production flow, rebalancing during production is very necessary work.
(4) Sewing is the most labor-intensive compared to the other three links of apparel manufacturing process that almost all the operations are conducted manually since the degree of automation is still low in the operation where non-rigid materials are processed (Zoumponos and Aspragathos 2008). This condition could be improved through polishing operators' skills and improving machine intelligence.
(5) Online and final inspections of semifinished and finished garments in sewing process are very important for quality control in apparel manufacturing process. Additionally, quality control can be realized by prediction and inspection of fabric performance.

Intelligent technologies are helpful for layout optimization to reduce fabric cost, and also these methods could be employed in sequencing sewing orders, balancing and rebalancing of sewing operations for shortening makespan. On aspect of quality control, advanced robotic sewing techniques could partly take place of labor force and at the same time improve product quality, where big data could benefit from its ability of prediction.

2.3 Finishing and Packing

Finishing or called ironing is to straighten the garments for packing. The packing process concerns a series of actions: sort, pile, and pack. In real production, finishing and packing sections are usually placed at one location. AI-related technology improvement related to these two sections lies in replacing manual operations with quick and accurate robotic operations by realization of robotic fabric folding for finishing and establishment of low-cost automatic system for packing.

3 Applications of the AI-Related Approaches

AI, also machine intelligence, combines a wide variety of advanced technologies to give machines an ability to learn, adapt, make decisions, and display new behaviors. Considerable efforts have been made for application of AI methodologies to handle some decision-making problems of apparel manufacturing in fashion industry in the last two decades. In this section, the existing literature was reviewed and the AI technologies used in the literature were further analyzed.

3.1 Literature Review Analysis

In this part, 59 articles are classified in terms of the year of publication, the type of journal, the authors, regions and institutions, and the application areas of apparel manufacturing process.

3.1.1 Distribution of Articles by Years

Figure 2 shows the distribution of articles published by year from 2000 to 2017. During the covered period, the number of published AI-related articles in apparel manufacturing process ranged from a low of 0 article in 2017 to the high of 8 in 2009, and the average number in each year is around 3. Researchers showed a great passion for AI study around the year of 2009, while the last decade saw a slow

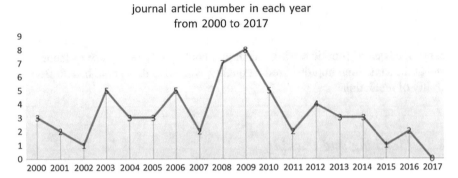

Fig. 2 Distribution of articles on AI in apparel manufacturing process for each year

drop. Till 2017, there was no publication which may be due to reasons like journal publication schedules and process of paper review and revision. This phenomenon indicated that in the forthcoming future as more new AI technologies emerge and the trend to realize the automation of apparel manufacturing, much more efforts should be made under the intelligent manufacturing concept.

3.1.2 Distribution of Articles by Journals

A total of 33 different journals from different disciplines are included in this literature review. As noted in Fig. 3, the top five most relevant categories included journals that focus on these issues are Expert Systems with Applications, International Journal of Clothing Science and Technology, International Journal of Production Research, Textile Research Journal, and The International Journal of Advanced Manufacturing Technology. Out of these, Expert Systems with Applications is the greatest source of articles, sharing 12% (7 out of 59), while the second largest source is The International Journal of Advanced Manufacturing Technology, with 10% (6) and followed by other three journals with the equal amount of three articles for each. Among them, Textile Research Journal and International Journal of Clothing Science and Technology are two scientific journals professional in the field of textile and clothing. The AI-related publications indicated the potential of AIs in accelerating the development of fashion industry. For the other three journals, Expert Systems with Applications is a refereed international journal that focuses on expert and intelligent systems technology and the related applications, International Journal of Production Research is a leading journal reporting researches regarding manufacturing, production, and operations management, and The International Journal of Advanced Manufacturing Technology provides an outstanding forum for applications-based researches relevant to manufacturing processes, machines, and process integration. It shows that study on AI technologies in apparel manufacturing does contribute to other manufacturing areas.

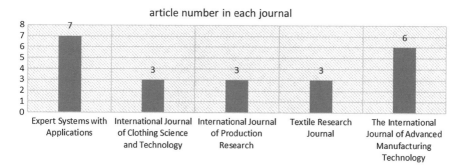

Fig. 3 Distribution of articles on AI in apparel manufacturing process by journals (the top 5)

3.1.3 Distribution of Articles by Authors, Regions, and Institutions

A total of 29 authors from 11 different regions were identified from the identified articles. Figure 4 presents the authors whose publication number on the related topic is 3 and above. The four authors, Guo Z.X., Hui C.L., Lee C.K.H., and Wong W.K., all come from Hong Kong have a bigger amount of the contribution with the article number of 6, 3, 4, and 12, respectively. Through this classification, it can be seen that Hong Kong had the largest percentage of publications as is shown in Fig. 5, followed by Greece and Taiwan with the percentages of 10 and 9%.

These contributing authors are from in total 18 institutions. Figure 6 shows the institutions which produced at least two articles. Institute of Textiles and Clothing, the Hong Kong Polytechnic University, ranks the top first with 28 articles and followed by University of Patras, Greece, and the University of Hong Kong, Hong Kong, with the numbers of 6 and 5 separately.

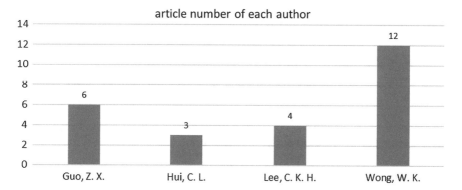

Fig. 4 Distribution of articles on AI in apparel manufacturing process by contributing authors (the top 4)

region of articles

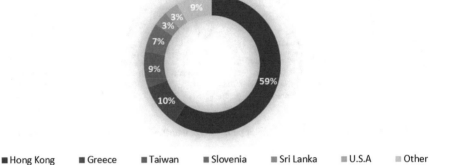

■ Hong Kong   ■ Greece   ■ Taiwan   ■ Slovenia   ■ Sri Lanka   ▥ U.S.A   ▥ Other

Fig. 5 Distribution of articles on AI in apparel manufacturing process by regions

institutions with more than 2 articles

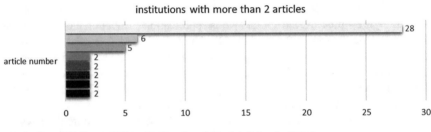

Institute of Textiles and Clothing, The Hong Kong Polytechnic University, Hong Kong

Department of Mechanical Engineering and Aeronautics, University of Patras, Greece

Department of Industrial and Manufacturing Systems Engineering, The University of Hong Kong, Hong Kong

Mura, European Fashion Design, Slovenia

Department of Textiles, Faculty of Mechanical Engineering, University of Maribor, Institute of Textiles, Slovenia

Department of Textile and Clothing Technology, University of Moratuwa, Sri Lanka

Department of Fashion Design and Management, Tainan University of Technology, Taiwan

Department of Management, College of Business Administration, University of Puerto Rico - Rio Piedras, USA

Fig. 6 Distribution of articles on AI in apparel manufacturing process by institutions (with over two articles)

3.1.4 Distribution of Articles by Application Areas

Soft computing techniques are mostly used for tackling different manufacturing issues during the past two decades. Likewise, the three techniques, Genetic Algorithm (GA), Fuzzy Logic (FL), and Neural Network (NN), have been used abundantly in apparel manufacturing. The AI methodologies are classified into five categories in this section: (1) GA, (2) FL, (3) NN, (4) other AI methodologies, and (5) hybrid intelligence (HI). Other AI methodologies contain Ant Colony Optimization (ACO) and Simulated Annealing (SA), and the HI category denotes hybrid methodologies which are the combinations of multiple AI techniques. The distribution of publication numbers according to AI methodologies by application areas is presented in Table 2.

EA, FL, NN, and the combination of EA and FL are the top 4 AIs with 26, 9, 9, and 5 publications and account for 44, 15, 15, and 8% separately. Obviously, EA is the most widely used AI exiting in the most amount of the articles. According to the literature, EA is the mainly used AI method for lay planning, SCS, and SALB. It can be indicated from the data that single intelligence like EA, FL, and NN and the combination of these three methodologies were the most popular technologies in the existing literature and single AI was preferred to HIs.

3.2 AI-Related Approaches Analysis

The above short analysis of the existing literature provides a brief introduction to the development state of AIs in apparel manufacturing, and then this part is to give the detailed analysis of these practical applications.

3.2.1 Genetic Algorithms

Both GA and Evolution Strategies (ES) belong to the EA category. A total number of 35 times EA intelligence occur in the existing literature, where GA accounts for the absolute largest percentage of 85% with an occurrence number of 30.

GA, developed by Holland (1992), is an adaptive heuristic search method based on natural selection. By exchanging intergroup information, GAs have a high local and global searching ability generating a whole population instead of one possible solution in order to avoid getting stuck on a local optimum. GA is a popular tool used for modeling and solving complex discrete optimization problems.

An increasing number of GA applications have appeared in the literature regarding apparel production scheduling and sequencing.

GA/EA-related approaches are used in scheduling problems including lay planning, marker optimization, facility layout, and SALB.

One of the first attempts of applying EA in lay planning of cutting process was made by Martens (2004). He developed a pair of GAs based on two alternative Integer Programming (IP) models with objectives of minimizing waste and overproduction, keeping costs at an acceptable level, and minimizing computation time. Later, EA-based methodologies were proposed to minimize marker number so as to reduce marker-making cost, fabric cost, as well as setup cost (Fister et al. 2008, 2010; Wong and Leung 2008; Abeysooriya and Fernando 2012a, b).

Yeung LHW proposed a hybrid method combining GA and the "Lowest-Fit-Left-Aligned" algorithm (LFLA) with which maker making was converted into a simple

Table 2 Distribution of AIs in different application areas of apparel manufacturing

Issue		AI approach									
		SI					HI				
		Main SI			Other SI		Main HI				Other HI
		EA	FL	NN	ACO	SA	EA+FL	EA+NN	FL+NN	EA+FL+NN	EA+SA
COP	COP	Lay planning	6									1
		Marker making	1			1	1		1			
SCS		6	1				3				
Roll sequencing		1									
Sewing order planning		2									
Line design	Layout design	1									
	SALB	8	1			2	1				
Product inspection			1	3				1			
Defects prediction				6							
Fabric handling			3						2	1	
Order planning		1									
Resource allocation			2								
Production monitoring			1				1				
Total		26	9	9	1	3	5	2	2	1	1
		44%	15%	15%	2%	5%	8%	3%	3%	2%	2%
		44			4		10				1
		48					11				

permutation problem and the optimal results can be obtained in a reasonably short period of time (Yeung and Tang 2003).

Lin MT addressed a single-row machine layout problem using a hierarchical order-based GA to quickly identify an optimal layout in a U-shaped sewing line for effective moving distance for cut-pieces with lower production costs (Lin 2009).

SALB has drawn the most attention of EAs researchers with the purpose of minimizing makespan and idle time. The application of EAs induced the increase of production efficiency. GAs were used to solve balancing problem of flexible assembly lines which allow flexible operation assignment, where one operation can be assigned to multiple workstations and multiple operations can be assigned to the same workstation (Hajri-Gabouj 2003; Guo et al. 2008b, c, 2009). For others, Wong W.K. adopted a GA to solve the SALB problem in UPS by periodically readjusting operator assignment. It was concluded that the optimal number of task skills each sewing operator should possess was three (Wong 2006a). In Chen et al. (2012), Mok et al. (2013), GAs were developed for automatic job allocations which can allocate workload among machines as evenly as possible for different labor skill levels. Zeng X. investigated the operator allocation problems with jobs sharing and operator revisiting for balance control of a complicated hybrid assembly line by EA (Zeng et al. 2012).

GAs are used in sequencing problems like order planning, SCS, and spreading roll sequencing.

In a Just-In-Time (JIT) manufacturing environment, jobs that are completed early must be held in a finished goods inventory til their delivery date, while jobs that are completed after their due dates may incur penalty costs. Therefore, Wong W.K. used a GA to achieve an ideal schedule in which all jobs are finished exactly on the assigned due dates (Wong and Chan 2001). To generate the optimal order scheduling solution in real-life make-to-order production with various uncertainties, Guo Z.X. proposed a GA-based approach with the objectives of maximizing the total satisfaction level of all orders and minimizing their total throughput time (Guo et al. 2008a).

Wong W.K. and his team published several articles applying GA to realize SCS balance and minimize the makespan in the period of 2000 and 2007. He applied GA to determine the number of spreading table installed in computerized fabric-cutting system (Wong et al. 2000b). GA afterward brought shorter completion time, higher machine utilization, and improvement of cut-piece fulfillment rates for a traditional manual system (Wong et al. 2005a), a computerized system (Wong et al. 2000a), and a manual–computerized system (Wong 2003b), with the JIT philosophy (Wong et al. 2001) or considering different types of existing uncertainties (Wong 2003a).

Wong W.K. proposed a GA to realize a Real-time Segmentation Rescheduling (RSR) of marker making, spreading, cutting, and bundling in dynamic apparel manufacturing environment (Wong et al. 2005b).

For fault-tolerant fabric-cutting schedule in a manual fabric-cutting manufacturing system, GAs were used in Wong (2006b) and Kwong et al. (2006) to satisfy resource-competing requests from downstream operating units/to minimize the makespan in a JIT production environment, and also a GA optimization procedure aiming to minimize the makespan was proposed in Mok et al. (2007).

In Hui et al. (2000), GA was applied to determine the optimal sequence of fabric rolls for each cutting lay-in fabric spreading with the purpose of fabric saving.

3.2.2 Fuzzy Logic

FL (Zadeh 1965, 1996) is an analysis method used to solve uncertain and imprecise problems. It has become a successful modeling tool for complex problems which can be controlled by humans but difficult to define precisely. Imprecise information as those resulting from inexact measurements or gaining from imperfectly codifying expert knowledge can be incorporated into a fuzzy modeling. Therefore, FL systems possess characteristics of simplicity and flexibility.

Decision making in manufacturing requires considering multitude of uncertainties. Variations in human operator performance, inaccuracies of process equipment, and volatility of environmental conditions are but just a few types of these uncertainties. Fuzzy tools have shown to perform just as or better than other soft approaches in decision making with uncertainties.

FL helps to deal with uncertainties in real-life environment and is used for robotic handling in sewing operation, SCS, SALB, resource allocation, and QC.

Automatic fabric handling uses FL to handle the varieties during conducting robot guiding non-rigid fabrics toward sewing. Koustoumpardis P.N. designed and developed a hierarchical robot control system including a fuzzy decision mechanism where the fuzzy rules and the membership functions were determined according to the experts' knowledge, combined with a neuro-controller to regulate the tensional force applied to the fabric during the robotized sewing process (Koustoumpardis and Aspragathos 2003). Later, he investigated the robotized sewing of two plies of fabrics (Koustoumpardis and Aspragathos 2014). Zoumponos G.T. presented a robot end-effector path-planning algorithm based on FL for the robotic laying of fabrics on a worktable which possesses the characteristics of flexibility and low computational cost (Zoumponos and Aspragathos 2008). Visual servoing attracted researchers' attention that Zacharia P. developed a flexible automation system tolerating deformations that may appear during robot handling of fabrics due to buckling without the need for fabric rigidification based on FL (Zacharia et al. 2009). The extended research of the same author focused on operations with curved-edge fabrics and correcting the distortions presented during robot handling of fabrics (Zacharia et al. 2010). Then, Zoumponos GT presented a new fuzzy visual servoing strategy based on the knowledge of easily measured fabric shape features for the folding of rectangular fabric strips by robotic manipulators (Zoumponos and Aspragathos 2010).

FL could deal with various existing uncertainties, such as deviations between estimated processing time and actual processing time, machine breakdowns, late receipt of fabric rolls, insertion of rush orders, in a dynamic cutting room. It has been used for SCS balancing problem in a computerized cutting system (Wong 2003a), for fault-tolerant fabric-cutting schedule (Mok et al. 2007), or in a manual cutting system with JIT production concept (Kwong et al. 2006; Wong 2006b).

Regarding resource allocation, Hui P.L. proposed an FL-related approach to determine the right number of operators to be moved in and out of sewing sections to insure overtime balance (Hui et al. 2002). Based on the knowledge representation aptitude of FL, the mathematical model was relaxed to overcome the nonlinearity and the complexity using fuzzy penalty functions to handle task-operator-machine assignment problem with multi-level objectives (Hajri-Gabouj 2003). Lee C.K.H. used FL to provide the optimal solution for resource allocation determination according to expertise knowledge stored as fuzzy rules (Lee et al. 2014b).

In the QC area, FL was used to find the relationships between production process parameters and product quality as knowledge support for parameter settings of machinery resources (Lee et al. 2013). Shu M.H. developed fuzzy-weighting on distinct defect severity classes for defects assignment to monitor textile-stitching nonconformities conditions (Shu et al. 2014).

3.2.3 Neural Networks

NN was first introduced by McCulloch and Pitts (1943) which was inspired by both biological nervous systems and mathematical theories of learning and can be used as a universal function approximator that learns from observations (training samples). Such systems learn (progressively improve performance) to do tasks by considering examples, generally without task-specific programming. NN is a highly connected network of processing elements or neurons arranged in three or more layers. It can demonstrate capabilities, such as prediction, pattern recognition, generalization, fault tolerance, and high-speed information processing. It learns from historical examples known as training data. During the training, the weights connecting the neurons present in different layers are optimized in such a fashion that the error signal reduces in each iterative step. Once sufficiently trained, NN can be used to solve unknown instances of the problem. NNs are one of the fastest most flexible classifiers used for fault detection due to their nonparametric nature and ability to describe complex decision regions.

In apparel manufacturing, NN is mainly used in QC, as well as maker making and robotic fabric handling.

NN is the most popular and the only AI method used for sewing performance prediction and inspection. An early attempt was in Kim et al. (2004), where NN was utilized to carry out system identification of the sewing machine motion system with a Brushless Direct Current (BLDC) motor for the purpose of accurate system control. Likewise, in Lin (2004), the NN predictive model on fabric and sewing thread optimization was developed for a commercial sewing machine equipped with BLDC motor. Hui C.L. investigated the use of Extended Normalized Radial Basis Function (ENRBF) NN to predict the sewing performance (Hui and Ng 2005). NN was used for studying the correlations between fabric mechanical properties and the seam appearance quality to construct the model for predicting seam. Compared with the traditional proposed Back Propagation (BP) NN in terms of prediction errors, the ENRBF Neural Network was proved to be better. Pavlinic D.Z. investigated the impact of fabric mechanical properties on the quality of seam appearance, which was defined by seam puckering and workpiece flotation. In the study, machine learning methods including regression trees (CART) and k-Nearest Neighbors (k-NN) were used to construct the model for predicting seam quality where the latter method is appreciated (Pavlinić et al. 2006). Pavlinic D.Z. developed a subjective evaluation system of garment appearance quality by studying the correlations between fabric mechanical parameters and the grade of garment appearance quality using k-NN (Pavlinic and Gersak 2009). Hui C.L. proposed BPNN for the prediction of fabric sewing performance which was classified in terms of four main factors: pucker, needle damage, fabric distortion, and fabric overfeeding based on 21 physical and mechanical properties (Hui et al. 2007). Afterward, Hui C.L. used BPNN with weight decay technique to predict the seam performance of commercial woven fabrics measured by the ratings of three indices, including seam puckering, seam flotation, and seam efficiency (Hui and Ng 2009). Mak K.L. presented an objective evaluation method based on NN to grade seam pucker with a high accuracy rate (Mak and Li 2008). Specially for knitted fabric, in Yuen et al. (2009a), four characteristic variables were collected and input into a BPNN to classify the sample images. Similarly, Yuen C.W.M. used a three-layer BPNN to deal with intelligent classification of fabric stitches or seams of semifinished and finished garments (Yuen et al. 2009b). The classifier with nine characteristic variables was proved outperformed those with five or seven variables and either BP or Radial Basis (RB) NN technique was effective where the recognition rate of BP NN was 100%.

To solve the marker-making problem, Wong W.K. constructed an irregular object packing approach where a learning vector quantization NN was developed by a set of examples inspired by experienced packing planners to diminish the size of a search space by dividing the objects into three classes (Wong and Guo 2010).

Regarding robotized sewing, Koustoumpardis P.N. developed a neuro-controller to regulate the tensional force applied to single piece fabric. In the further research (Koustoumpardis and Aspragathos 2003), he controlled the force applied by the robotic manipulator with NN in order to join two pieces of fabrics (Koustoumpardis and Aspragathos 2014). Zacharia P. dealt with the curved edges of real cloth parts by NN to learn from the information obtained from the fabrics used for training and then respond to new fabrics (Zacharia et al. 2009).

3.2.4   Other AI Methodologies

Apart from the three most widely used AIs, ACO had been mentioned once and SA three times in the literature.

ACO is an optimization algorithm inspired by collective behaviors of ant colonies when they search for food. Taking advantage of the information ants released, the optimal path/solution can be obtained. ACO is broadly used to solve computational path-finding problems.

In the apparel manufacturing related literature, ACO was used to select the appropriate combination of cutting patterns in solving cutting layout problem.

Yang C.L. combined an ACO and an IP to solve cutting layout problem where the ACO was applied for selecting the appropriate combination of cutting patterns (Yang et al. 2011).

SA is commonly the first algorithm that using an explicit strategy to escape from local optima. The main idea is to allow solutions with worse quality to escape from local optima. The probability of doing such an escape is decreased during the search process. Up to date, SA is still an interest of further studies, as tool for tackling many optimization problems and as component of other algorithms.

For the apparel manufacturing process, SA is mainly used for production planning problems, like marker making and SALB.

Javanshir H. applied SA to solve the two-dimensional cutting stock problem and calculate the optimum length of fabric rolls in order to reduce the cutting waste (Javanshir et al. 2010).

Ruiz-Torres A.J. investigated the flow shop scheduling problem with operations flexibility in purpose of minimizing the number of tardy jobs/the completion time of all the jobs. SA was employed in the proposed two-stage neighborhood search improvement method (Ruiz-Torres et al. 2010, 2011).

3.2.5 Hybrid Intelligence

For decision making in real-world production, it is usual that multiple production objectives need to be considered and achieved simultaneously. To achieve the multi-objective optimal solution, hybrid intelligence for utilizing an integration of AIs' advantages makes sense.

We have the five combinations here: EA+FL, EA+NN, FL+NN, EA+FL+NN, and EA+SA in 11 out of total 59 articles from the literature.

EA(GA)+FL In SCS area, GA was used with fuzzy due times to generate fault-tolerant manual fabric-cutting schedules in a JIT production environment (Kwong et al. 2006; Wong 2006b; Mok et al. 2007). This method could effectively satisfy the demand for downstream production units and reduce production costs through reducing operator idle time.

For SALB problem, Hajri-Gabouj S. developed a genetic optimization algorithm with fuzzy penalty relaxation to realize flexible assignment in sewing lines with multi-level objectives (Hajri-Gabouj 2003).

EA+NN Wong W.K. constructed an irregular object packing approach that integrated a grid approximation-based representation, a learning vector quantization NN, a heuristic placement strategy, and an integer representation-based $(\mu + \lambda)$ ES to increase the usability of the stock sheet for solving the marker-making problem (Wong and Guo 2010).

Yuen C.W.M. presented a hybrid model integrating GA and BPNN for knitted garment defects classification (Yuen et al. 2009a).

FL+NN FL and NN were combined and used in fabric handling.

Koustoumpardis P.N. proposed a hierarchical robot control system including a fuzzy decision mechanism where the fuzzy rules and the membership functions were determined according to the experts' knowledge, combined with a neuro-controller to regulate the tensional force applied to the fabric during the robotized sewing process (Koustoumpardis and Aspragathos 2003). Later, he applied the method in the robotized sewing of two-ply fabrics (Koustoumpardis and Aspragathos 2014).

EA(GA)+FL+NN Zacharia P. presented the design and tune of the adaptive neuro-fuzzy inference systems (ANFIS) for robot guiding fabrics with curved edges toward sewing which is powerful to deal with uncertainty embedded in the curved edges of real cloth parts and new fabric pieces that had not been used for the training process (Zacharia et al. 2009).

EA+SA For the issue of COP, M'Hallah R. combined lay planning and maker making into a single problem whose objective was to minimize fabric length. It was solved using constructive heuristics and metaheuristics: stochastic local improvement method (SA), global improvement method (GA), and hybrid approach. These designed approaches were proved to present a trade-off between efficacy and efficiency (M'Hallah and Bouziri 2016).

3.3 Conclusion

Some conclusions based on the analysis above are offered as follows:

(1) Although a considerable number of application areas were applied by AI methodologies by which contributions were made in apparel manufacturing, there are still some sections need more attention, like cutting route planning, sorting and bundling in cutting, ironing, finishing and packing, material transportation and management in sewing, inventory management. The exploitation of using AI methodologies may be helpful in improving the performance of these areas and, in turn, benefitting all sectors in fashion industry. Following are four points to be improved:

Innovative Optimized COP Concept
In traditional cutting department, lay plan and marker making are handled as two successive separate processes. Visibly, scientific estimated marker-making results are the proofs of good lay plan. That means it is to combine these two issues to propose an innovative optimized COP concept that can generate the optimal COP scenario satisfying certain constraints. Logic of the new COP concept is shown in Fig. 7 that after lay planning, marker efficiency can be figured out based on result of marker making, according to which adjustments could be done to produce optimized lay plan and then evaluate again. The loop repeats until the requirement is reached and then output the optimal scenario.

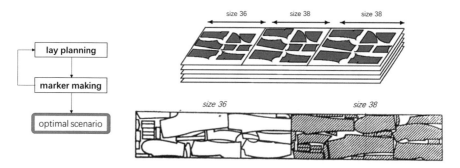

Fig. 7 Logic of the innovative optimized COP concept

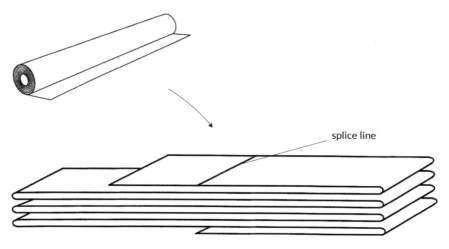

Fig. 8 Optimization of fabric rolls sequence

Fabric Rolls Arrangement

On aspect of fabric spreading, optimal selection of fabric width corresponding to different apparel type and optimal arrangement of fabric rolls with different lengths for lays are good means to save fabric. Figure 8 is a sketch showing the optimization of fabric rolls arrangement.

Cutting Route Optimization

It makes sense to figure out the optimal cutting route which will contribute to both makespan minimization and resources utilization improvement. AIs, for instance, ACO could be applied to find the shortest path. In Fig. 9, it is shown that for the same marker different options of cutting paths can be offered.

Cut-Piece Sorting and Bundling

As shown in Fig. 10, the increasing diversity and variety of the patterns make it more complicated to carry out the sorting and bundling work. The classification method with shortest operation time and of high accuracy is in need.

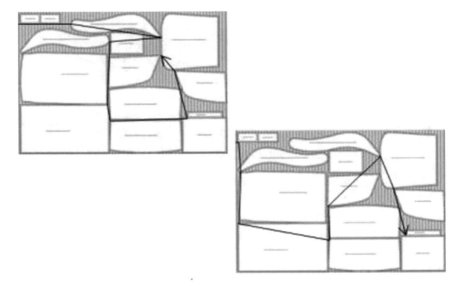

Fig. 9 Cutting path planning in cutting process

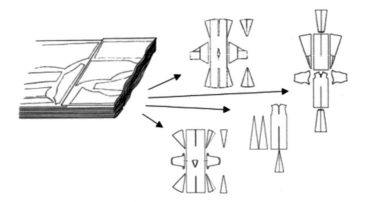

Fig. 10 Complex sorting and bundling process after cutting operation

(2) The study of the whole apparel manufacturing process has attracted some atten-
    tion of AI researchers. However, still most of the applications of AIs are confined
    to certain single function areas. As AI methodologies can offer advantages in
    managing information flow, which is an important factor in enhancing the inte-
    gration of these processes, so the full use of these methodologies will make it
    easier for promotion of communication and relevant information sharing among
    all sections, in order to realize the true integration of all sections in the whole
    apparel manufacturing process.

(3) Single intelligence always shows its disadvantage in certain individual area
    and also has some short points, while a hybridization of intelligence could be

more powerful because the integration could bring supplements and improvements. The optimization, hybridization, and creation of AIs could be studied for establishing professional and proper AI technologies to provide effective solutions based on specific characteristics in apparel manufacturing such as the uncertainties and dynamic features (the variety of operators' skill level, operator absenteeism, machine breakdowns, unexpected orders, etc.).

(4) Parameter setting of an AI technique has a large effect on its performance. Due to the insufficiency of theoretical foundation of AI techniques, parameter setting usually depends on experience or the trial-and-error method in previous studies and it is not determined accurately and specific for different problems. The way of how to set AIs-related parameters is a good issue to be considered next.

(5) Only a few of the proposed AI methods were employed in real-life apparel manufacturing, and this is because the vast majority of the methods are with a lack of evaluation in actual production. More practice should be conducted in actual manufacturing process to evaluate the proposed technologies.

4 New Perspectives

As the mass customization trend emerging in modern fashion industry, apparel manufacturing moves from large volume to small volume with greater product variability, different to make-to-stock, no inventory of finished or semifinished goods is available to buffer demand fluctuations, and therefore the survival of apparel company depends on its ability of quick response to order requirements and its flexibility in production operations which raises the demand of advanced technologies for production prediction, management, and control.

Big data uses inductive statistics and concepts from nonlinear system identification to infer laws (regressions, nonlinear relationships, and causal effects) from large sets of data with low information density so as to reveal relationships and dependencies or to perform predictions of outcomes and behaviors. At present, advanced AI tools become more widespread due to the power and affordability of automation and computerization. Those specific advanced AIs could be employed for value extraction from the data.

Based on the big data concept, in accordance with the research status from the literature, research trends and the guidance of information-oriented industrial production are proposed. Potential and applicable solutions for guidance of future work are suggested as follows.

Realization of the Synergy Effects Across Whole Production
The Internet of Things (IoT) is the network of physical devices and other items embedded with electronics, software, sensors, actuators, and network connectivity which enable these objects to connect and exchange data. The IoT allows objects to be sensed or controlled remotely across existing network infrastructure.

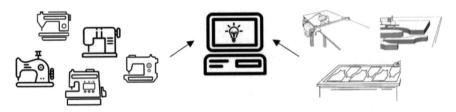

Fig. 11 Central computing system in apparel manufacturing

Processes in the production chain have interactions with its previous and subsequent processes. The problems considering integrated multi-processes of apparel manufacturing should be further studied. Data exchange of different apparel manufacturing processes and cloud computation could, the concept of Industry 4.0, tend to better benefit the overall manufacturing performance.

Production capacity of cutting process should match that of sewing process which means the cutting department offers adequate cutting-pieces on time satisfying the time and quantity constraints. Both of these two departments need to have a good understanding of each other's progress and keep in pace for the smooth of the whole production. As shown in Fig. 11, the information communication could be built between cutting process and sewing process. Data collected in these two physically separated places is processed and analyzed in computer for realizing the synergy effects across garment production.

Application of RFID Technology

One focus of the smart manufacturing is creating manufacturing intelligence from real-time data to support timely and accurate decision making. With the recent application of new technologies, such as RFID, real-time production information has become available for decision support in manufacturing system. Yuan, L. offered real-time tracking and monitoring of materials flowing in manufacturing processes which contribute to quick response for adjustments to the ever-changing production environment (Yuan and Tang 2017).

Cost is the core concern in manufacturing, and it is necessary to establish efficient resource management systems concerning material transportation and quantity, labor quantity and working time, machine quantity and working time, inventory and overproduction. It could be easier to master a manufacturing system with the RFID devices-equipped resources.

Increase in Specialization and Adaptability

More details should be put into study for the new demands of modern manufacturing or the basic characteristics of apparel manufacturing itself.

Mass customization leads to various pattern characteristics like color, size, material, and style/type. In face of the mass customization trend, to group patterns of diverse garment sizes or types (as shown in Fig. 12) and cut out article pieces together, then trim individually could be a nice way to gain efficiency. That is to say, patterns are classified into groups before cutting. And the way of pattern group division could

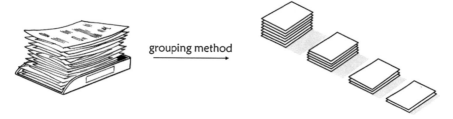

Fig. 12 Pattern grouping for mass customization

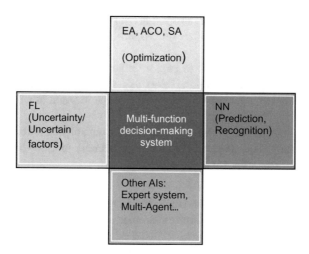

Fig. 13 Extension of multi-AI methodology

be studied. The grouping criterion could be abstracted from the pattern characteristics and based on pattern deviation.

As apparel manufacturing environment is fuzzy and dynamic, AI technologies could be optimized to be applicable to apparel manufacturing for uncertainty, subjectivity, ambiguity, and vagueness embedded like the dynamic and fuzzy factors of operator's skill level, operator absenteeism, deviations between estimated processing time and actual processing time, machine failures, interruption of job processing owing to quality problems, late receipt of materials fabric rolls, insertion of rush orders.

Establishment of Multi-function Decision-Making System

An integrated solution of multi-techniques is capable to pursue superior results where big data is used to train AIs. Figure 13 shows a multi-function decision-making system gathering abilities from different AIs.

It is also necessary for further research to improve the efficiency and effectiveness of existing AI algorithm or to create a new one or to integrate more practical algorithms.

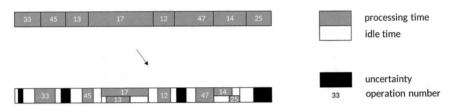

Fig. 14 Real-time adjusting SCS scenario update model

One application is to figure out the optimal cutting route for minimizing operation time and raising resources utilization rate, where ACO could be applied to find the shortest path.

In addition, it is also possible to explore other types of AI, for example: tabu search, Particle Swarm Optimization (PSO), expert system, and multi-agent system, which has already been proved of great performance in some other manufacturing processes for the field of apparel.

Advanced AIs for robotic operation optimization with objectives of high accuracy, extensive capability, high speed and low cost, especially in labor-intensive sewing operation, are the key point for automation.

Production Simulation

Simulation technology can help evaluate the proposed approach by modeling the scenario applied with the approach in a simpler and low-cost way. Here, big data could be used to produce input data for model adjustment. There have been studies on simulations of sewing lines, and it shows great performance to apply simulation method to production management. Thus, extending applications would be establishing the simulation model, as shown in Fig. 14, for easy and quick cutting capacity estimation serving as the base to work out the timely adjusted smooth spreading and cutting workflow.

References

Abeysooriya RP, Fernando TG (2012) Hybrid approach to optimize cut order plan solutions in apparel manufacturing. Int J Inf Technol Commun Res 2

Abeysooriya RP, Fernando TGI (2012b) Canonical genetic algorithm to optimize cut order plan solutions in apparel manufacturing. J Emerg Trends Comput Inf Sci 3(2):150–154

Chen JC, Chen CC, Su LH, Wu HB, Sun CJ (2012) Assembly line balancing in garment industry. Expert Syst Appl 39(11):10073–10081

Fister I, Mernik M, Filipič B (2008) Optimization of markers in clothing industry. Eng Appl Artif Intel 21(4):669–678

Fister I, Mernik M, Filipič B (2010) A hybrid self-adaptive evolutionary algorithm for marker optimization in the clothing industry. Appl Soft Comput 10(2):409–422

Guo ZX, Wong WK, Leung SYS, Fan JT, Chan SF (2006) Mathematical model and genetic optimization for the job shop scheduling problem in a mixed-and multi-product assembly environment: a case study based on the apparel industry. Comput Ind Eng 50(3):202–219

Guo ZX, Wong WK, Leung SYS, Fan JT, Chan SF (2008a) Genetic optimization of order scheduling with multiple uncertainties. Expert Syst Appl 35(4):1788–1801

Guo ZX, Wong WK, Leung SYS, Fan JT, Chan SF (2008b) A genetic-algorithm-based optimization model for solving the flexible assembly line balancing problem with work sharing and workstation revisiting. IEEE Trans Syst Man Cybern C 38(2):218–228

Guo ZX, Wong WK, Leung SYS, Fan JT, Chan SF (2008c) A genetic-algorithm-based optimization model for scheduling flexible assembly lines. Int J Adv Manuf Technol 36(1–2):156–168

Guo ZX, Wong WK, Leung SYS, Fan JT (2009) Intelligent production control decision support system for flexible assembly lines. Expert Syst Appl 36(3):4268–4277

Guo ZX, Wong WK, Leung SYS, Li M (2011) Applications of artificial intelligence in the apparel industry: a review. Text Res J 81(18):1871–1892

Guo ZX, Ngai EWT, Yang C, Liang X (2015) An RFID-based intelligent decision support system architecture for production monitoring and scheduling in a distributed manufacturing environment. Int J Prod Econ 159:16–28

Hajri-Gabouj S (2003) A fuzzy genetic multiobjective optimization algorithm for a multilevel generalized assignment problem. IEEE Trans Syst Man Cybern C 33(2):214–224

Holland JH (1992) Adaptation in natural and artificial systems: an introductory analysis with applications to biology, control, and artificial intelligence. MIT Press

Hui CL, Ng SF (2005) A new approach for prediction of sewing performance of fabrics in apparel manufacturing using artificial neural networks. J Text Inst 96(6):401–405

Hui CL, Ng SF (2009) Predicting seam performance of commercial woven fabrics using multiple logarithm regression and artificial neural networks. Text Res J 79(18):1649–1657

Hui PC, Ng FS, Chan KC (2000) A study of the roll planning of fabric spreading using genetic algorithms. Int J Cloth Sci Tech 12(1):50–62

Hui PL, Chan KC, Yeung KW, Ng FF (2002) Fuzzy operator allocation for balance control of assembly lines in apparel manufacturing. IEEE Trans Eng Manag 49(2):173–180

Hui PC, Chan KC, Yeung KW, Ng FS (2007) Application of artificial neural networks to the prediction of sewing performance of fabrics. Int J Cloth Sci Tech 19(5):291–318

Javanshir H, Rezaei S, Najar SS, Ganji SS (2010) Two dimensional cutting stock management in fabric industries and optimizing the large object's length. J Res Rev Appl Sci 4(3):243–249

Kim IH, Fok S, Fregene K, Lee DH, Oh TS, Wang DW (2004) Neural network-based system identification and controller synthesis for an industrial sewing machine. Int J Control Autom 2(1):83–91

Koustoumpardis PN, Aspragathos NA (2003) Fuzzy logic decision mechanism combined with a neuro-controller for fabric tension in robotized sewing process. J Intell Rob Syst 36(1):65–88

Koustoumpardis PN, Aspragathos NA (2014) Intelligent hierarchical robot control for sewing fabrics. Robot Comput Integr Manuf 30(1):34–46

Kwong CK, Mok PY, Wong WK (2006) Determination of fault-tolerant fabric-cutting schedules in a just-in-time apparel manufacturing environment. Int J Prod Res 44(21):4465–4490

Lee CK, Choy KL, Ho GT, Law KM (2013) A RFID-based resource allocation system for garment manufacturing. Expert Syst Appl 40(2):784–799

Lee CK, Ho GT, Choy KL, Pang GK (2014a) A RFID-based recursive process mining system for quality assurance in the garment industry. Int J Prod Res 52(14):4216–4238

Lee CK, Choy KL, Law KM, Ho GT (2014b) Application of intelligent data management in resource allocation for effective operation of manufacturing systems. J Manuf Syst 33(3):412–422

Lee CK, Choy KL, Ho GT, Lam CH (2016) A slippery genetic algorithm-based process mining system for achieving better quality assurance in the garment industry. Expert Syst Appl 46:236–248

Lin TH (2004) Construction of predictive model on fabric and sewing thread optimization. J Text Eng 50(1):6–11

Lin MT (2009) The single-row machine layout problem in apparel manufacturing by hierarchical order-based genetic algorithm. Int J Cloth Sci Tech 21(1):31–43

Mak KL, Li W (2008) Objective evaluation of seam pucker on textiles by using self-organizing map. IAENG Int J Comput Sci 35(1)

Martens J (2004) Two genetic algorithms to solve a layout problem in the fashion industry. Eur J Oper Res 154(1):304–322

McCulloch WS, Pitts W (1943) A logical calculus of the ideas immanent in nervous activity. Bull Math Biophys 5(4):115–133

M'Hallah R, Bouziri A (2016) Heuristics for the combined cut order planning two-dimensional layout problem in the apparel industry. Int T Oper Res 23(1–2):321–353

Mok PY, Kwong CK, Wong WK (2007) Optimisation of fault-tolerant fabric-cutting schedules using genetic algorithms and fuzzy set theory. Eur J Oper Res 177(3):1876–1893

Mok PY, Cheung TY, Wong WK, Leung SY, Fan JT (2013) Intelligent production planning for complex garment manufacturing. J Intell Manuf 24(1):133–145

Pavlinic DZ, Gersak J (2009) Predicting garment appearance quality. Open Text J 2:29–38

Pavlinić DZ, Geršak J, Demšar J, Bratko I (2006) Predicting seam appearance quality. Text Res J 76(3):235–242

Ruiz-Torres AJ, Ablanedo-Rosas JH, Ho JC (2010) Minimizing the number of tardy jobs in the flowshop problem with operation and resource flexibility. Comput Oper Res 37(2):282–291

Ruiz-Torres AJ, Ho JC, Ablanedo-Rosas JH (2011) Makespan and workstation utilization minimization in a flowshop with operations flexibility. Omega 39(3):273–282

Shu MH, Chiu CC, Nguyen TL, Hsu BM (2014) A demerit-fuzzy rating system, monitoring scheme and classification for manufacturing processes. Expert Syst Appl 41(17):7878–7888

Wong WK (2003a) A fuzzy capacity-allocation model for computerised fabric-cutting systems. Int J Adv Manuf Technol 21(9):699–711

Wong WK (2003b) Optimisation of apparel manufacturing resource allocation using a generic optimised table-planning model. Int J Adv Manuf Technol 21(12):935–944

Wong WK, Chan CK (2001) An artificial intelligence method for planning the clothing manufacturing process. J Text Inst 92(2):168–178

Wong WK, Guo ZX (2010) A hybrid approach for packing irregular patterns using evolutionary strategies and neural network. Int J Prod Res 48(20):6061–6084

Wong WK, Leung SY (2008) Genetic optimization of fabric utilization in apparel manufacturing. Int J Prod Econ 114(1):376–387

Wong WK, Chan CK, Ip WH (2000a) Optimization of spreading and cutting sequencing model in garment manufacturing. Comput Ind 43(1):1–10

Wong WK, Chan CK, Ip WH (2000b) Effects of spreading-table quantities on the spreading-table planning of computerized fabric-cutting system. Res J Text Appar 4(1):25–36

Wong WK, Chan CK, Ip WH (2001) A hybrid flowshop scheduling model for apparel manufacture. Int J Cloth Sci Tech 13(2):115–131

Wong WK, Chan CK, Kwong CK, Mok PY, Ip WH (2005a) Optimization of manual fabric-cutting process in apparel manufacture using genetic algorithms. Int J Adv Manuf Technol 27(1–2):152–158

Wong WK, Leung SY, Au KF (2005b) Real-time GA-based rescheduling approach for the pre-sewing stage of an apparel manufacturing process. Int J Adv Manuf Technol 25(1–2):180–188

Wong WK, Mok PY, Leung SY (2006a) Developing a genetic optimisation approach to balance an apparel assembly line. Int J Adv Manuf Technol 28(3–4):387–394

Wong WK, Kwong CK, Mok PY, Ip WH (2006b) Genetic optimization of JIT operation schedules for fabric-cutting process in apparel manufacture. J Intell Manuf 17(3):341–354

Yang CL, Huang RH, Huang HL (2011) Elucidating a layout problem in the fashion industry by using an ant optimisation approach. Prod Plan Control 22(3):248–256

Yeung LH, Tang WK (2003) A hybrid genetic approach for garment cutting in the clothing industry. IEEE Trans Ind Electron 50(3):449–455

Yuan L, Tang W (2017) A compact broadband UHF RFID tag antenna for metallic objects. In: Applied computational electromagnetics society symposium (ACES), Suzhou, 28 September 2017

Yuen CW, Wong WK, Qian SQ, Chan LK, Fung EH (2009a) A hybrid model using genetic algorithm and neural network for classifying garment defects. Expert Syst Appl 36(2):2037–2047

Yuen CW, Wong WK, Qian SQ, Fan DD, Chan LK, Fung EH (2009b) Fabric stitching inspection using segmented window technique and BP neural network. Text Res J 79(1):24–35

Zacharia PT (2010) An adaptive neuro-fuzzy inference system for robot handling fabrics with curved edges towards sewing. J Intell Rob Syst 58(3–4):193–209

Zacharia P, Aspragathos N, Mariolis I, Dermatas E (2009) A robotic system based on fuzzy visual servoing for handling flexible sheets lying on a table. Ind Robot 36(5):489–496

Zadeh LA (1965) Fuzzy sets. Inform Control 8:338–353

Zadeh LA (1966) Fuzzy sets. Fuzzy sets, fuzzy logic, and fuzzy systems: selected papers by Lotfi A Zadeh, pp 394–432

Zeng X, Wong WK, Leung SY (2012) An operator allocation optimization model for balancing control of the hybrid assembly lines using Pareto utility discrete differential evolution algorithm. Comput Oper Res 39(5):1145–1159

Zoumponos GT, Aspragathos NA (2008) Fuzzy logic path planning for the robotic placement of fabrics on a work table. Robot Comput Integr Manuf 24(2):174–186

Zoumponos GT, Aspragathos NA (2010) A fuzzy strategy for the robotic folding of fabrics with machine vision feedback. Ind Robot 37(3):302–308

A Discrete Event Simulation Model with Genetic Algorithm Optimisation for Customised Textile Production Scheduling

Brahmadeep and Sébastien Thomassey

Abstract This chapter aims to explain the methodology of the production schedule optimisation for the automatic manufacturing of customised textile products. The data involved in this manufacturing process are huge and constitute many parameters and constraints. The proposed system could be divided into two main modules, the optimisation model and the production floor model. Indeed, the complexity of this scenario demands a hybrid model which involves a combination of an optimisation model (genetic algorithm model) and a production simulation model (discrete event simulation) with a robust link (ActiveX/OLE Automation Server). The system forms a complex synchronised loop which replicates and improves the production schedule in process till the best results are achieved. The expected impacts are to have on-time shipment, increased productivity and profitability with the implementation of lean tools. Indeed, the implementation of this model is very vast. This would permit the use of a powerful discrete event model with an optimisation algorithm which gives numerous possibilities from manufacturing scheduling to the global supply chain, distribution and logistics planning and optimisation.

1 Introduction

The customer needs are becoming more personalised which require industries to produce in small lot sizes, short lead time and with unique specifications. This leads to a high variety of products and consequently a huge number of information to manage in production. Thus, the industries face challenges to still reduce the manufacturing costs, increase the efficiency of the present inventory and resources and manage more data. This trend of mass customisation and personalised production leads to fundamental changes in the material flow, plant layout and work organisation (Acaccia et al. 1999). They have to rely on highly flexible production system

Brahmadeep · S. Thomassey (✉)
ENSAIT, GEMTEX, 2 allée Louise et Victor Champier, BP
30329, 59056 Roubaix Cedex 1, France
e-mail: sebastien.thomassey@ensait.fr

© Springer Nature Singapore Pte Ltd. 2018
S. Thomassey and X. Zeng (eds.), *Artificial Intelligence for Fashion Industry in the Big Data Era*, Springer Series in Fashion Business,
https://doi.org/10.1007/978-981-13-0080-6_8

coupled with efficient and intelligent planning management, and overall supervision and decision systems. This could be very challenging for some companies such as textile companies which are commonly faced with large batches, long process and set-up times.

The flexibility, generally reached with autonomous and independent workshops, in such a production system requires a robust planning and scheduling tool to achieve maximum utilisation, efficiency, on-time shipment and profit. A system with the most optimised and suitable production tools will be privileged in order to obtain the best product at an acceptable cost and on-time.

The high variety of products and different specifications involves many constraints in productions especially when the machine and human resources are limited and a little or no flexible (e.g. because of long set-up processes or high qualified tasks). In this framework, it becomes difficult to rely on formal or analytic techniques to model and optimise the production process. Indeed, to deal with the large number of parameters, nonlinear influences, time and casual dependencies and stochastic parameters, simulation modelling coupled with meta-heuristic optimisation appears as the more suitable technique.

The aim of this chapter is to explain the methodology of the production schedule optimisation for garment manufacturing of customised textile products using simulation–optimisation models. The expected impacts are to have on-time shipment, increased productivity and profitability with the implementation of lean tools.

The main challenges to reach this aim are to discuss the requirement for the optimisation of the automatic production based on the Key Performance Indicators (KPIs), to define the production simulation logic with parameters and constraints, to develop a hybrid model which integrates the production simulation model with the optimisation model and establish a robust link between both, to simulate the hybrid model to find the best possible optimised production schedule, to analyse the results obtained from the simulations based on the defined fitness function and finally to simulate the best optimised schedule and compare the results.

This chapter is structured as follows. In Sect. 2, an overview of the existing literature and case studies related to this work is presented. In Sect. 3, the methodology of this work is described along with the detailed definition of the problem, parameters, constraints and the development of the hybrid simulation model. In Sect. 4, the details of the experimental work executed and results obtained are formulated, analysed and compared. An overview of the future scopes and development of this work is explained with a global conclusion in Sect. 5.

2 State of the Art

2.1 Simulation in Manufacturing and Textile Production

There are many factors which contribute to the use of the simulation in the manufacturing. The advantages include analysing the manufacturing processes, ease of use, flexibility, ability to model dynamic and stochastic nature of production systems, ability to test various scenarios of production, layout design, logistics, material handling, which lead us to have a close insight into the performance of a manufacturing company (Chen and Harlock 1999; Chen et al. 2002; Wanga et al. 2011; Ünal et al. 2009; Ekren and Omek 2008).

The application of simulation in manufacturing can be seen in the case of some modern complex concepts like backward online change scheduling in production where the insight is the WIP management, lean processes and reduction of idle times explained in Kim and Choi (2014). Also, Edis and Ornek (2009) explained a case analysis of lot streaming in job shops with transportation queue disciplines where there are multi-products and sub-lots. It considered the idling and availability, set-up time of the machines, load–unload and trip times of the transporter, weekly demands and job routings. The manufacturing units having the concepts of Make-to-Order utilise the advantages of the simulation-based models. The proposed model by Ôzbayrak et al. (2007) is developed using a system dynamics approach where the operations performed within a supply chain are a function of a great number of key variables which have strong interrelationships and a key step towards the optimised performance.

The textile and apparel manufacturing units are indeed very complex systems because of the number of production parameters and high variety of references. Nowadays, the demand is also moving towards more personalised products which would give rise to small lot sizes, short lead times and unique specifications. To compete in this changing environment, the producer has to improve their layout, productivity, reactivity, flexibility and quality. Each layout problem has its own unique characteristics and the objectives (Nahmias 2005).

Flexibility is also an essential requirement in order to respond to shorter product life cycles, low to medium production volumes, changing demand patterns and a higher variety of product models and options (Bukchin et al. 2002). The flexible nature of the simulation model to find the feasible solutions for the problems such as planning and scheduling of the fabric production orders, to achieve the shipment on time, maximising the loom utilisation, labour requirement, etc., is described by Chen and Harlock (1999). These results demonstrated the flexible nature of the simulation model, and this application can be easily extended to other textile sectors like apparel production and customisation, supply chain.

The application of simulation in apparel and fashion is very vast. It is used to meet the challenges inherent in producing a greater range of products, but at the same time trying to keep the economy-of-scale benefits of mass production (Sepulveda and Akin 2004). The simulation tools are used to link all processes of a flexible production system to have an intelligent system to analyse clothing manufacturing. The set-ups are analysed by computer simulation in order to verify the actual potentials of the facilities with a foresight to find the most effective plans for long-run operations by adaptive exploitation of flexibility at different levels with effective schedules and to find the best facility design to meet the growing demands which are personalised, quick response and total quality with reduced lead times (Ünal et al. 2009; Acaccia et al. 2003; Greasley 2008; Zülch et al. 2011). The application of simulation for the forecasting of the sourcing processes of a retailer and a manufacturer that enables to quantify the impact of the forecast errors on the financial and supply performances which are linked with the production plan, inventory level, service level and profit margin of the manufacturer is demonstrated by Thomassey (2010).

These works verify that the simulation models are very powerful and efficient to enable the companies to achieve their goals in this competitive world by analysing different scenarios by simulation models and later implementing the proposed organisational changes. The use of simulation in the optimisation of the production planning and scheduling requires a robust optimisation tool. Meta-heuristics, evolutionary algorithms such as genetic algorithm are considered as one of the most suitable solutions for such scheduling problems. The next subsection presents some literature and case studies which illustrate the application of this optimisation tool.

2.2 Scheduling and Optimisation by Genetic Algorithm

In the 1950s and the 1960s, several computer scientists independently studied evolutionary systems with the idea that evolution could be used as an optimisation tool for engineering problems. The idea in all these systems was to evolve a population of candidate solutions to a given problem, using operators inspired by natural genetic variation and natural selection. These forms of systems were known to be the evolutionary algorithms (Mitchell 1996).

In the domain of artificial intelligence, an evolutionary algorithm is a subset of evolutionary computation, a generic population-based meta-heuristic optimisation algorithm. An evolutionary algorithm uses mechanisms inspired by biological evolution, such as reproduction, mutation, recombination and selection. Candidate solutions to the optimisation problem play the role of individuals in a population, and the fitness function determines the quality of the solutions. Evolution of the population then takes place after the repeated application of the above operators. Bäck (1996) explained that the artificial evolution describes a process involving individual evolutionary algorithms; EAs are individual components that participate in an artificial evolution. There are different techniques or types by which an evolutionary algorithm model can work, one of which is genetic algorithm.

Genetic algorithms (GAs) were invented by John Holland in the 1960s and were developed by Holland and his students and colleagues at the University of Michigan in the 1960s and the 1970s. In contrast to evolution strategies and evolutionary programming, the original goal was not to design algorithms to solve specific problems, but rather to formally study the phenomenon of adaptation as it occurs in nature and to develop ways in which the mechanisms of natural adaptation might be imported into computer systems (Holland 1975).

GA is a family of computational models inspired by evolution. These algorithms encode a potential solution to a specific problem on a simple chromosome like data structure and apply recombination operators to these structures so as to preserve critical information. Genetic algorithms are often viewed as function optimisers although the range of problems to which genetic algorithms have been applied is quite broad.

An implementation of a genetic algorithm begins with a population of typically random chromosomes. One then evaluates these structures and allocates reproductive opportunities in such a way that those chromosomes, which represent a better solution to the target problem, are given more chances to reproduce than those chromosomes which are poorer solutions. The goodness of a solution is typically defined with respect to the current population (Michalewicz 1992).

The application of GA in the scheduling is very vast. Some case studies related are discussed in the following paragraphs ranging from the production scheduling to the routing problems.

In Palencia and Delgadillo (2012), the authors investigated a sequencing of the assembly line as a resource constraint scheduling problem. The proposed solution method was a classical genetic algorithm which was adapted to handle the constraints of the production system. The algorithm is embedded into a computer application that calculated the weekly schedule. The implementation has brought dramatic changes. The percentage of on-time deliveries rose from 65 to 85% in about one and a half years.

GA is implemented to optimise the total cost and service level for just-in-time distribution in a supply chain by Farahani and Elahipanah (2008). A similar case of JIT where the design of the supply chain for the auto parts was optimised by the GA operations is described by Shi et al. (2014). Further, the authors of Peralta et al. (2014) demonstrated the GA implementation to optimise the production, cost and energy with dynamic return flow. This case corresponds to the water resource optimisation where a simulation–optimisation tool was used to maximise water provided from sources, maximise hydropower production and minimise operation costs of transporting water from sources to destinations. Vallada and Ruiz (2011) proposed a genetic algorithm for the parallel machine scheduling problem with sequence-dependent set-up times with the objective to minimise the makespan.

Various hybrid models which implement GA operations were used to solve the problems related to resource constraint production in Gonçalves et al. (2008). Toledo et al. (2014) demonstrated a multi-level GA model in which the optimisation problems in a soft drink-manufacturing unit such as lot scheduling and sizing issues were solved. In case of the flexible manufacturing units where the parameters are mainly the job-routing flexibility and sequence of operations on each machine, Giovanni and Pezzella (2010) used an improved GA to optimise the job shop scheduling and routing of the operations. These steps can be implemented in the large set of distributed-and-flexible scheduling problems. Li et al. (1998) proposed a GA approach to address the problem of earliness and tardiness production scheduling and planning. The authors proposed methods which included lot-size considerations and capacity balancing. The optimisation of such parameters ensured that the manufacturer responded to the changing market requirements quickly and fulfilled the needs of the customers.

The simulation–optimisation demonstrated by Azadivar and Tompkins (1999) showed the application GA with qualitative variables and structural model changes. The optimum response sought is a function of design and operation characteristics of the system such as the type of machines to use, dispatching rules, sequence of processing operations. The lot numbers in production are unique numbers, where the permutation A is implemented such as in travelling salesman problem (TSP). A case described by Salomon et al. (1997) demonstrated the application of TSP in the discrete lot-sizing and scheduling problem with sequence-dependent set-up costs and set-up times. A similar case where the scheduling problems on a single machine in a sequence-dependent set-up environment are optimised by the TSP-GA tools is described by Bigras et al. (2008).

The literature verifies genetic algorithm as one of the optimisation tools suitable for the manufacturing sector. Indeed, GA has been successfully implemented for specific problems of the textile apparel industry such as: cutting order planning (Leung and Wong 2013), marker planning (Wong et al. 2013a), fabric spreading and cutting scheduling (Wong et al. 2013b) or apparel production order planning (Guo et al. 2013). But to optimise a scheduling problem in case of complex systems like the manufacturing of customised textile products where the parameters and constraints are huge in numbers, a hybrid model is required. The proposed solution is a multi-level synchronised model which integrates a discrete event simulation to deal with the specificities and complexities of the manufacturing units and products variety with an optimisation engine. The next subsection presents some literature and case studies related to the application of hybrid models.

2.3 Hybrid Model Integrating a Discrete Event Simulation Model with an Optimisation Model

The complexity that is involved in the scheduling of the manufacturing processes inspired the development of multi-level hybrid GA models which integrate discrete event simulation tools like the *Arena Simulation* which can make the global optimisation tool very robust and powerful (Liang and Yao 2008). As per Dias et al. (2011), the *Arena Simulation* is one of the most popular and powerful simulation software in its domain. The modules in the software play an important role in modelling an automated manufacturing process or with an involvement of scheduling, queuing and waiting. It is integrated with VBA coding which enhances the possibilities of the simulation model (Wanga et al. 2011). Li et al. (2011) proposed a methodology for a scheduling problem and to optimise the makespan and the total tardiness of the whole production. It demonstrated a similar hybrid model which linked the simulation software *Arena* and a fuzzy optimisation algorithm optimisation model. The results were compared and concluded that the advantages of the hybrid model are better than the traditional methods.

   Korytkowski et al. (2013) developed a methodology to optimise the dispatching scheduling using *Arena Simulation* and integrating it with GA model in *MATLAB*. This technique aided to maximise the performance of a complex manufacturing system with a large number of different products, along with an overtime that changes with a mix of different process types, including assembly and disassembly operations and with different types of internal and external disturbances. The methodology to connect and build a robust link between *Arena* and a GA program in *MATLAB* is demonstrated by Ouabiba et al. (2001) and Liang and Yao (2008).

   The above literature and case studies suggest to proceed with the GA operations while integrating it with a discrete event simulation software with a robust link to achieve maximum efficiency in the optimisation process. This solution emerges as the more efficient way to deal with the complexity, the huge variety, the constraints, etc.

3 Methodology

3.1 Description of the Manufacturing Unit

The manufacturing unit considered in this study is an automated textile product factory for customised production. This customisation involves basic operational attributes like logo, button types, fabric flaps or attachments. The production is scheduled according to the shipment date of the orders.

   The company deals with the high variety of specifications, small lot sizes and short lead times. The production section is composed of specialised dedicated machines for the required customised operation. Each machine has sections which are more specific to the operations involved while each section has slots where the work is actually executed (a number of slots represent the capacity of the section). There are also operators and technicians who are responsible for the supervision of the processes per shift. The factory works full time on weekdays (Monday to Friday).

   The factory material flow starts from the warehouse where the raw material (unfinished products before customisation) is stored. The warehouse stock is maintained by a push flow, which is as per the forecasted demand. The orders in the warehouse consist of information related to the product requirements, shipment dates and customer information.

   The automatic customisation process starts from the warehouse where the required products are transferred via the conveyor systems to the machines as per the schedule based on the shipment dates of the orders (Fig. 1).

   The distribution logic of the unfinished products to the machines is based on the machine parameters since they are ultimately distributed to the machines. The

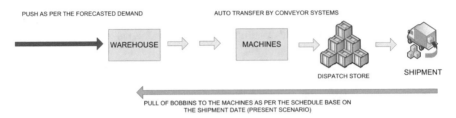

Fig. 1 Present flow of materials in the factory

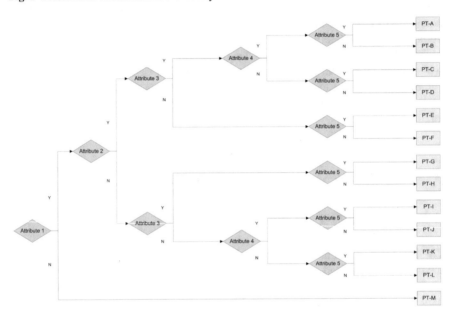

Fig. 2 Product categorisation based on the machine parameters

products are categorised on the basis of machine parameters from PT-A to PT-M; see Fig. 2. This logic confirms that a particular category can go to a particular machine and to a particular section for the process. The machine set-up does not change for the shift of one category of product to another because the categories are already determined as per the machines. Later the finished products are automatically transferred to the packaging section and are kept ready for final shipment in a dispatch store.

As per the information from the concerned textile manufacturing unit, a first comparison between manual and automatic scenarios of production was analysed for a week on the basics of two performance indicators that are global cost of production and total delay in shipment (Brahmadeep and Thomassey 2014). The aim of this preliminary study was to quantify the profits generated with the automation of different tasks (mainly transfer between machines). It was found that, in case of the manual (standard) production process, the total delay in shipment of the products is lower than the automatic (enhanced) process but the global cost is higher than

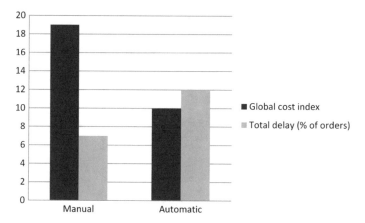

Fig. 3 Comparison between manual and automatic production

the automatic process. Simple comparative results of the scenarios are mentioned in Fig. 3 where an index value of 10 is taken for the automatic process and the value of manual production is derived from that index. In case of the delays, total percentage of the delayed orders for the shipment is considered. This result can be explained by the ability to the manual method to deal with urgencies and unplanned events (such as breakdowns) whereas the automatic process strictly follows the expected scheduling.

The aim of the concerned production unit is to reduce the cost which can be achieved by the automatic production, but it involves delays in shipment hence a possibility of losing customers. The solution is to optimise the automatic production schedule to achieve zero delays in shipment of the products resulting in minimum global cost of production and staying competitive in the competitive market.

3.2 Production Parameters, Constraints and Simulation Logic

3.2.1 Description of the Simulation and Optimisation Tools

As per the literature previously discussed in Sect. 2.3, to simulate the production scenario, *Arena Simulation* is selected. It is a powerful simulation software to model manufacturing units like the case discussed in this study. The integration of the VBA coding further enhances the modelling parameters and scopes which are required for recording multiple personalised results.

As discussed in Sect. 3.1, there is a need to optimise the production schedule. For the optimisation of such production system, an evolutionary algorithm like the GA is used specifically for a permutation-based GA. Indeed, the order numbers in the

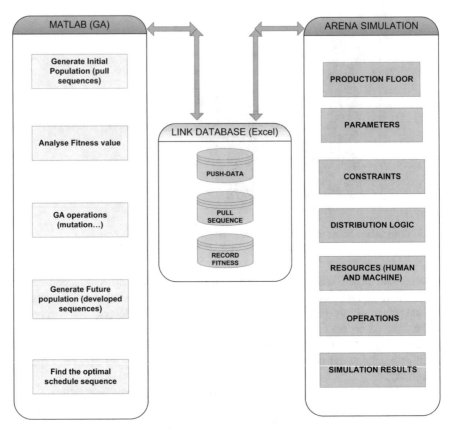

Fig. 4 Hybrid simulation model with respective properties and parameters

factory are unique and the scheduling sequence must have all the unique numbers to represent the same. For the system to run at maximum efficiency with minimum idle time and delivery of the products on time, the scheduling of the inputs to the system must be optimised to the best possible level. It will select the best sequences and will reject the rest based on a **fitness function** (minimise delays and maximise on-time/before-time deliveries, see Sect. 3.2.2).

The implementation of the model can only be possible with a robust link between these tools. A database in the centre is used to link and generate a hybrid model. Figure 4 describes each part of the hybrid model with respect to the parameters, processes and operations.

Briefly, *MATLAB* platform is used for the GA optimisation model which is linked to the event simulation software *Arena*. A link is established between these two tools by *Microsoft Excel* to form a hybrid model.

3.2.2 Assignment of the Fitness Function

The aim of the fitness function in the optimisation algorithm is to calculate the total delay of an order sequence and also calculate the time value of the on-time or before-time shipped orders on the same schedule. This allows to obtain a schedule with more safety as the results from the optimisation will eliminate delays and also increase the value of before-time shipment of orders. The increase in before-time value would give the manufacturing unit for flexibility and minimise resources/inventory and time to perform more activities if required.

The fitness function for a schedule considered for the optimisation is presented as follows:

$$f(N) = \sum_{i=1}^{N} f(i)$$

where

$$f(i) = \begin{cases} 1 + |d_i|, & d_i < 0 \\ \\ \frac{1}{(1+d_i)}, & d_i \geq 0 \end{cases}$$

N   Number of orders in the schedule;
d_i $(t_s - t_a)$, delay or on-time or before-time of orders in hours;
t_s Target shipment date and time;
t_a Actual (simulated) shipment date and time.

The next step is to implement the hybrid model which would integrate the production simulation model with the GA optimisation tool discussed before.

3.2.3 Hybrid Model with Robust and Synchronised Link

The hybrid simulation model for the optimisation using GA consists of various synchronised loops (GA iterations) which makes a robust system to make an efficient decision-making.

The process starts with the execution of the *MATLAB* script (optimisation model) which generates the initial population of sequences (1 population = 100 sequences = 100 schedulings) and writes it on an *Excel* sheet, which also sends a signal to the *Arena* discrete event simulation (production model) to start the event (production) simulation for each sequence (100 replications for 100 sequences or 1 population).

In the meantime, the optimisation model waits for the results for each sequence which is obtained from the *Arena* Simulation via *Excel* sheets. With all the required data obtained, GA operations are executed in the optimisation model and a new generation of population is generated and recorded on the sheets. With the execution of each loop, an improved set of sequences is generated. These iterations continue till the optimal results are obtained, i.e. minimum fitness value.

The event production model (*Arena*) receives push-data from another file which consists of various parameters like number of orders, size of orders, target delivery date, quality. The link and signalling between these models are managed by an automation server, COM/OLE ActiveX. This automation server provided real-time and quick response data linking and signalling. A detailed demonstration of flow diagram is provided in Fig. 5.

4   Experimentation and Results

For the optimisation experiments, one-week data are taken which consisted of 471 orders with all the parameters (for push input). The pull order sequences are taken as per the shipment dates for the present scenario (before optimisation), and later in the optimisation process, these are generated by the GA model.

Each iteration of the GA model generates a population of 100 sequences. Each sequence consists of 471 unique numbers between 1 and 471 which represent the order numbers for the pull. The GA model generates a new population at each iteration based on the objective of the fitness function which is minimisation in this case (Fig. 5).

4.1   Results Obtained from Before Optimisation

The delays and on-time shipment of orders obtained before the optimisation, namely the orders are scheduled according to their shipment date, are illustrated in Fig. 6. The x-axis represents the actual 471 shipped orders for one week, and the y-axis represents the value of d_i per order where:

d_i $(t_s - t_a)$, delay or on-time or before-time of orders in hours;
t_s Target shipment date and time;
t_a Actual (simulated) shipment date and time.

(zero value $=$ on-time, positive value $=$ before-time and negative value $=$ delay).

The total aggregate of shipment delays of the all the orders is found to be 43.16 h, and before-time value is 74,703.85 h.

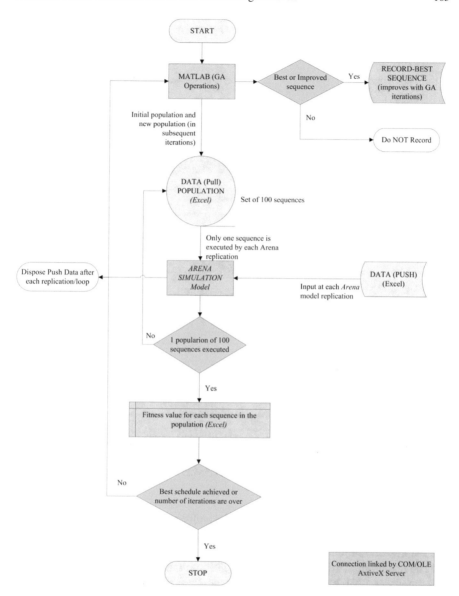

Fig. 5 Operation sequence of the hybrid simulation model

4.2 GA Hybrid Model Optimisation Results

The same simulation of the production of the 471 orders is performed with the developed hybrid model. The optimisation results are based on the fitness function formulated in Sect. 3.2.2 and are illustrated in Fig. 7. The x-axis is represented by

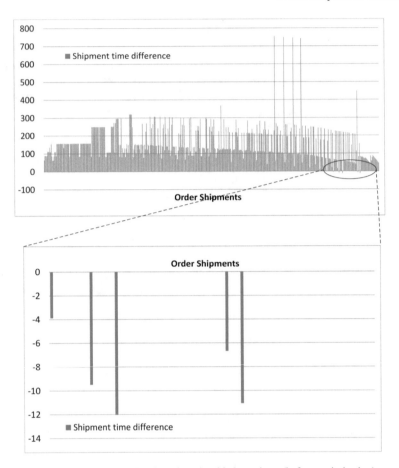

Fig. 6 Shipments as per the schedule based on the shipment dates (before optimisation)

the number of GA iterations (model loops), and y-axis represents the fitness value $f(N)$. As a reminder, the objective of this function is to have zero delay and maximise the before-time shipments.

The results show that the fitness value got minimised to the global minima after 721 iterations (721 loops in the hybrid model). It can be seen in Fig. 7 that the algorithm enables a fast and strong reduction of the fitness function from a very high value (from 864 to 5) which corresponds to zero delay schedulings (from 864 to 5) and then deeper optimises the scheduling in order to increase before-time shipments (from iteration 50 to iteration 721). The overall best sequence for the pull is recorded and stored. The next step is to verify the best sequence achieved and compare the results.

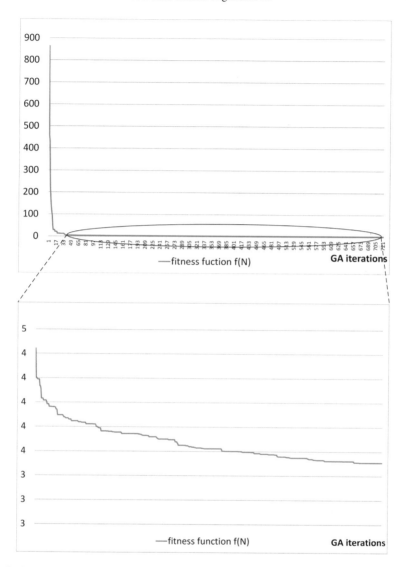

Fig. 7 GA optimisation based on the minimisation of the fitness function

4.3 Results Obtained from the Best Sequence by GA Hybrid Model

With the same simulation presented in Sect. 4.1 and the best sequence derived in the GA optimisation, the results (delays and on-time shipment of orders) obtained are illustrated in Fig. 8.

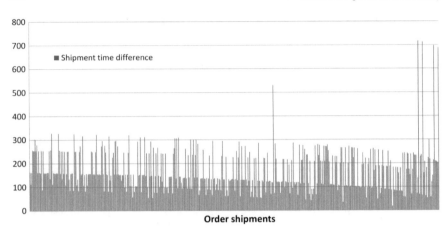

Fig. 8 Shipments as per the schedule based on the best sequence derived after the optimisation

Table 1 Result comparison

Basis of production schedule	Total accumulated delay (h)	Total accumulated before-time (h)
Shipment dates (%)	43.16	74,703.85
Optimised schedule	0	80,617.51

The total aggregate of shipment delays of the all the orders is found to be 0 (zero), and before-time value is 80,617.51 h.

4.4 Discussion

The optimised schedule when compared with the shipment dates-based scheduling shows 'zero' delays, and further, the before-time value is also improved (Table 1). This justifies the need for the optimisation and the application of the hybrid model. Consequently, the automated process with optimised production schedule would result in no delays with minimum costs.

It is also important to observe that the optimisation process is able to find a schedule with zero delay very quickly, namely around 50 iterations. The improvement of before-time shipments, which ensures more robustness against unexpected events, requires more computational efforts.

5 Conclusion and Scope

In this work, a methodology is developed for the optimisation of the production scheduling in an automated manufacturing unit dedicated to customised textile products. This scenario presents a considerable degree of detail and complexity. Thus, a hybrid simulation model is developed with a robust link which integrates the discrete event simulation and an optimisation model using GA algorithm. The link between the two modules relies on ActiveX/OLE Automation Server technology. This approach is designed to cope with the huge number of production parameters and data involved.

The automation of the manufacturing unit with a production scheduling based on the shipment dates of the orders involves lower cost than the manual process but more delays. Thus, the objective in this case study is to derive a production schedule which ensures 'zero' delay and maximise the before-time shipments. The increase in before-time value would give the manufacturing unit the flexibility, to minimise resources/inventory and time to perform more activities if required. From the results, it obviously appears that the optimised schedule enables a significant reduction of the delay, i.e. 'zero' delay and a higher before-time shipment value. These results also verify the robustness in synchronisation between the modules of the hybrid model.

Finally, the scope of this methodology is very vast. This hybrid model is flexible in nature which enables to change the parameters at ease. This would permit the use of a powerful discrete event model with an optimisation algorithm which gives numerous possibilities from manufacturing scheduling to the global supply chain, retail distribution, and logistics planning and optimisation, hence staying competitive in the global market.

References

Acaccia GM, Conte M, Maina D, Michelini RC, Molfino R (1999) Integrated manufacture of high standing dresses for customized satisfaction. In: Globalisation of manufacturing in the digital communication era, pp 511–523

Acaccia GM, Conte M, Maina D, Michelini RC (2003) Computer simulation aids for the intelligent manufacture of quality clothing. Comput Ind 50:71–84

Azadivar F, Tompkins G (1999) Simulation optimization with qualitative variables and structural model changes. Eur J Oper Res 113:169–182

Bäck T (1996) Evolutionary algorithms in theory and practice: evolution strategies, evolutionary programming, genetic algorithms. Oxford University Press, Oxford

Bigras LP, Gamache M, Savard G (2008) The time-dependent traveling salesman problem and single machine scheduling problems with sequence dependent setup times. Discret Optim 5:685–699

Brahmadeep, Thomassey S (2014) A simulation based comparison: manual and automatic distribution setup in a textile yarn rewinding unit of a yarn dyeing factory. In: Simulation modelling practice and theory, vol 45, pp 80–90

Bukchin J, Dar-El EM, Rubinovitz J (2002) Mixed model assembly line design in a make-to-order environment. Comput Ind Eng 41(4):405–421

Chen EJ, Lee YM, Selikson PL (2002) A simulation study of logistics activities in a chemical plant. Simul Model Pract Theory 10:235–245

Chen G, Harlock SC (1999) A computer simulation based scheduler for woven fabric production. Text Res J 69:431–439

Dias LS, Pereira G, Vik P, Oliveira JA (2011) Discrete simulation tools ranking: a commercial software packages comparison based on popularity. In: Proceedings of 9th annual industrial simulation conference. Industrial Simulation Conference, Venice

Edis RS, Ornek A (2009) Simulation analysis of lot streaming in job shops with transportation queue disciplines. Simul Model Pract Theory 17:442–453

Ekren BY, Ornek AM (2008) A simulation based experimental design to analyze factors affecting production flow time. Simul Model Pract Theory 16:278–293

Farahani RZ, Elahipanah M (2008) A genetic algorithm to optimize the total cost and service level for just-in-time distribution in a supply chain. Int J Prod Econ 111:229–243

Giovanni L, Pezzella F (2010) An improved genetic algorithm for the distributed and flexible job-shop scheduling problem. Eur J Oper Res 200:395–408

Gonçalves JF, Mendes JJM, Resende MGC (2008) A genetic algorithm for the resource constrained multi-project scheduling problem. Eur J Oper Res 189:1171–1190

Greasley A (2008) Using simulation for facility design: a case study. Simul Model Pract Theory 16:670–677

Guo ZX, Wong WK, Leung SYS, Fan JT, Chan SF (2013) Optimizing apparel production order planning scheduling using genetic algorithms. In: Leung S, Guo ZX, Wong WK (eds) Optimizing decision making in the apparel supply chain using artificial intelligence (AI): from production to retail. Woodhead Publishing, pp 55–80

Holland JH (1975) Adaptation in natural and artificial systems. University of Michigan Press, Michigan. Second edition (1992). The MIT Press, Massachusetts

Kim T, Choi BK (2014) Production system-based simulation for backward on-line job change scheduling. Simul Model Pract Theory 40:12–27

Korytkowski P, Wisniewski T, Rymaszewski S (2013) An evolutionary simulation-based optimization approach for dispatching scheduling. Simul Model Pract Theory 35:69–85

Leung S, Wong WK (2013) Optimizing cut order planning in apparel production using evolutionary strategies. In: Leung S, Guo ZX, Wong WK (eds) Optimizing decision making in the apparel supply chain using artificial intelligence (AI): from production to retail. Woodhead Publishing, pp 81–105

Li X, Chehade H, Yalaoui F, Amodeo L (2011) A new method coupling simulation and a hybrid metaheuristic to solve a multiobjective hybrid flowshop scheduling problem. In: Proceedings of the 7th EUSFLAT-LFA conference. Aix-les-Bains, France

Li Y, Ip W, Wang DW (1998) Genetic algorithm approach to earliness and tardiness production. Int J Prod Econ 54:65–76

Liang S, Yao X (2008) Multi-level modeling for hybrid manufacturing systems using Arena and Matlab. In: International workshop on modelling, simulation and optimization. Hong-Kong, China

Michalewicz Z (1992) Genetic algorithms + data structures = evolutionary programs. Springer, New York

Mitchell M (1996) An introduction to genetic algorithm. The MIT Press, Massachusetts

Nahmias S (2005) Production and operations analysis. McGraw-Hill, New York

Ouabiba M, Mebarki N, Castagna P (2001) Couplage entre des methodes d'optimisation iteratives et des modeles de simulation a evenements discrets. In: 3e Conférence Francophone de MOdélisation et SIMulation, Conception, Analyse et Gestion des Systèmes Industriels. Conference MOSIM, Troyes, France

Ôzbayrak M, Papadopoulou TC, Akgun M (2007) Systems dynamics modelling of a manufacturing supply chain system. Simul Model Pract Theory 15:1338–1355

Palencia AER, Delgadillo GEM (2012) A computer application for a bus body assembly line using genetic algorithms. Int J Prod Econ 140:431–438

Peralta RC, Forghani A, Fayad H (2014) Multiobjective genetic algorithm conjunctive use optimization for production, cost, and energy with dynamic return flow. J Hydrol 511:776–785

Salomon M, Solomon MM, Wassenhove LNV, Dumas Y (1997) Solving the discrete lotsizing and scheduling problem with sequence dependent set-up costs and set-up times using the Travelling Salesman Problem with time windows. Eur J Oper Res 100:494–513

Sepulveda JA, Akin HM (2004) Modelling a garment manufacturer's cash flow using object-oriented simulation. In: Ingalls RG, Rossetti MD, Smith JS, Peters BA (eds) Winter simulation conference, 2004, Proceedings of the winter simulation, vol 2. Association for Computing Machinery, New York, pp 121–128

Shi W, Shang WJ, Liu Z, Zuo X (2014) Optimal design of the auto parts supply chain for JIT operations- Sequential bifurcation factor screening and multi-response surface methodology. Eur J Oper Res 236:664–676

Thomassey S (2010) Sales forecasts in clothing industry: the key success factor of the supply chain management. Int J Prod Econ 128:470–483

Toledo CFM, Oliveira L, Pereira RF, França PM, Morabito R (2014) A genetic algorithm mathematical programming approach to solve a two-level soft drink production problem. Comput Oper Res 48:40–52

Ünal C, Tunali S, Güner M (2009) Evaluation of alternative line configurations in apparel industry using simulation. Text Res J 79:908–916

Vallada E, Ruiz R (2011) A genetic algorithm for the unrelated parallel machine scheduling problem with sequence dependent setup times. Eur J Oper Res 211:612–622

Wanga J, Chang Q, Xiao G, Wang N, Li S (2011) Data driven production modeling and simulation of complex automobile general assembly plant. Comput Ind 62:765–775

Wong WK, Wang XX, Guo ZX (2013a) Optimizing marker planning in apparel production using evolutionary strategies and neural networks. In: Leung S, Guo ZX, Wong WK (eds) Optimizing decision making in the apparel supply chain using artificial intelligence (AI): from production to retail. Woodhead Publishing, pp 106–131

Wong WK, Kwong CK, Mok PY, Ip WH (2013b) Optimizing fabric spreading and cutting schedules in apparel production using genetic algorithms and fuzzy set theory. In: Leung S, Guo ZX, Wong WK (eds) Optimizing decision making in the apparel supply chain using artificial intelligence (AI): from production to retail. Woodhead Publishing, pp 132–152

Zülch G, Koruca HI, Börkircher M (2011) Simulation-supported change process for product customization: a case study in a garment company. Comput Ind 62:568–577

An Intelligent Fashion Replenishment System Based on Data Analytics and Expert Judgment

Roberta Sirovich, Giuseppe Craparotta and Elena Marocco

Abstract Retail stock allocation is crucial but challenging. The authors developed an innovative solution, successfully tested in the context of high-end fashion: collaboration between artificial intelligence and human intuition. Each week, stores are assigned a budget based on current stock levels versus potential sales, and offered to "spend" this budget with an initial data-driven recommendation on which SKU/sizes order and release. Each store manager is then given a time window, so she can modify the proposal while respecting budget constraints; and finally, the artificial intelligence optimally allocates available stock to requests based on the expected likelihood of sale minus cost of logistics, subject to management-defined constraints. Our test showed how this system outperformed the control group of stores, relying on a traditional head office-driven allocation without direct human input. The retailer boosted sales, demand cover, and stock rotation performance: an estimated 1M EUR margin/month positive impact. Moreover, the new system improved store managers morale through non-monetary incentive-driven empowerment.

Keywords Retail · Artificial intelligence · Constrained optimization
Forecasting · Dynamic markets · Innovation · Luxury · Fashion

1 Introduction

Inventory allocation is crucial to retail, even more so to those verticals whose sales are difficult to accurately predict. Case in point is fashion retail: the SKU lifecycle is short, residual end-of-life values are either very low or zero, SKU performance is volatile, and the unit values at stake are high. Inventory allocation has to match finite stock resources with unknown demands, to optimally accelerate rotation, avoid missed sales opportunities, and therefore ultimately increase profits. The topic is even

R. Sirovich · G. Craparotta (✉) · E. Marocco
Department of Mathematics, University of Torino, Turin, Italy
e-mail: giuseppe.craparotta@unito.it; gcraparo@unito.it

© Springer Nature Singapore Pte Ltd. 2018
S. Thomassey and X. Zeng (eds.), *Artificial Intelligence for Fashion Industry in the Big Data Era*, Springer Series in Fashion Business,
https://doi.org/10.1007/978-981-13-0080-6_9

more challenging in the case of high-end brands, due to the concentration of value in a small amount of pieces.

Inventory decisions influence sales in many ways: different assortments can be assigned to different stores, due to local trends and to the store dimension; the allocation on sizes can be fundamental for customer satisfaction. At the same time, the retailer has to avoid missing sales coming from stock-out as possible. The fragmented and highly local item/size nature of the demand made it difficult to accurately predict sales, creating challenges to centrally driven stock allocation process. A uniform stock allocation across stores and seasons can be perceived by the management as sub-optimal; however, a more differentiated allocation needs to catch each store peculiarities and to guarantee at the same time the optimal performance of the whole network.

The purpose of this work is to show the implementation and results of an original intelligent system able to improve sales by reducing stock-outs and missed opportunities, thanks to improvements in forecast accuracy, even at a high granular level such as store/SKU/size. This approach can leverage the store managers' insight and experience, by directly feeding their input into a new allocation process. Additionally, the proposed system enables the direct transhipment across stores, to correct any errors in the initial stock allocation, and therefore increases the liquidity of the store network.

With the new approach, inventory allocation becomes part of a retail-integrated management process, from merchandise planning and open-to-buy management, to in-season sales forecasting, to the design and measurement of pricing and promotion initiatives. This approach achieved multiple ambitious goals: reduce unsold stock, increase rotation, empower store managers, increase relevance to local consumer demand, and increase profits.

This article is structured as follows: Sect. 2 presents the literature review. The data sets used are detailed in Sect. 3; Sect. 4 presents methodology and process macro-steps: forecast, proposal, internal marketplace, allocation, and shipment optimization. Section 5 reviews the main results.

2 Literature Review

Multiple authors covered the various components of fashion inventory management: how to buy inventory before the season starts, how to allocate stock to stores, how to update the merchandise plan during the season.

A dynamic programming approach is often used to solve such problems: Agrawal and Smith (2013) formulate a dynamic stochastic optimization model to determine total order size and optimal inventory allocation across non-identical stores for each period. They use Bayesian inference to model partially correlated demands across stores and time periods, and then dynamic programming to find an optimal allocation. The authors of Fisher and Rajaram (2000) apply linear programming to estimate the season forecast for the chain by testing a subset of the chain. However, these approaches do not factor in any human input.

Macy's reported that considering stores as homogeneous is sub-optimal: differences in store demands create opportunities to increase relevance and therefore ultimately grow profits.

Van Donselaar et al. (2010) highlights the diversity of demand across stores and provides a framework to apply *micromerchandising*, a practice followed by a significant number of retailers: each store has a unique assortment, that maximizes its consumer appeal. The author attempts to infer store managers' behavior in order to adapt the automatic replenishment. However, in practice it is operationally challenging to manage a dynamic programming model across multiple time periods and multiple stores.

Attempts at developing a traditional store-level forecast are generally hinded by the idiosyncrasies of the fashion industry, that make it complex and characterized by unsatisfactory results. Thomassey (2010) shows how fashion sales are highly volatile and fragmented, due to multiple intrinsically noisy and volatile factors: seasonality, fashion trends, promotions, and others.

The allocation system is often crucial in the relation between the company and the single stores, and any changes in the replenishment process reflects the company thinking. There are two common alternatives to manage the store replenishment: some models centralize the shipment decisions to achieve the best result for the entire network of stores, and other models make shipment decisions based primarily on stores' input. Some company experimented both automated as well as non-automated allocation systems. In the non-automated system, store managers are allowed to freely request inventory, and central oversight is limited to applying simple stock availability constraints and manually prioritizing requests. This system is relatively manual, time-consuming and sometimes not yielding the desired sales performance. Moreover, store managers often attempt to order more than necessary in order to receive a number of pieces that is close to what they truly want (as it is pointed out in Cachon and Lariviere (1999) for a similar framework). When dealing with scarce resource in a competitive environment, as pointed out in Furuhata and Zhang (2018), an optimal (from the retailer's point of view) allocation can be reached giving the stores truth-inducing mechanisms. On the other side, fully automated system managed centrally can lead smaller stores to consistently carry a limited range, while larger stores systematically received the best picks from the available inventory.

Multiple authors explored the possibility to combine an automatic store forecast with human expert judgment to leverage the impartiality of the former with the expertise of the latter. Among these, Blattberg and Hoch (1990) concluded that the combined forecast generally outperforms automatic-only and expert-only forecast. In the case of high-end fashion, the strong influence of local weather and consumer preferences on sales, combined with relatively deep knowledge of individual customers that store managers have, led us to choose them as the key providers of human judgment.

Caro and Gallien (2010) implemented a similar model for Zara that combines a forecast with store managers' demands. The core model is stochastic and predicts SKU-store-level sales during a replenishment period as a function of expected demand, available inventory, and store stock-out policy. They then formulate a mixed-integer program that embeds a piecewise-linear approximation of the first model applied to every store in the network; the output store shipment quantities maximize overall predicted sales subject to inventory availability and other constraints. The model is based on weekly forecasts, best suited to fast-moving items (such as those of Zara), and is optimal for the particular merchandising policy in which SKUs with size-level stock-outs are removed from the shelves. Human judgment is applied by multiple experts working in a competitive environment where total stock is constrained. Store managers would tend to over-order if no balancing mechanism was provided, so Caro and Gallien introduced weights for the automatic forecast and store managers. Separately, the same authors Gallien et al. (2015) provided an implementation of an automatic allocation algorithm of initial shipments from the central warehouse, based on forecast updating and dynamic optimization.

We propose a replenishment model combining automatic forecast and stores' input. In the high-end fashion segment, multiple differences make the mentioned approach of (Caro and Gallien 2010) less directly appliable: a greater emphasis on increasing stock rotation and reducing slow-moving items to avoid a very significant cash depreciation; and the high item value makes transhipment across stores economically feasible. Therefore, we adapted Caro and Gallien's model, allowing for stock releases (negative shipments to a store) and creating an internal cross-store marketplace that allows store managers to trade their stock. We do not approximate the optimization model; thus, we obtain an integer programming problem that is easier and faster to solve.

Additionally, we introduce a store budget to address the different levels of stock and sales potential across the stores. In case the store manager was not able to provide his forecast, we provide an initial proposal that approximates human behavior and tries to capture the local tendencies.

Creating a sort of internal marketplace among stores, the proposed system allows to differentiate each store's offer, according to the real local demand. Finally, the system allows to implement a defragmentation mechanism to minimize store-level stock-outs, by optimally matching missing sizes to overstocked sizes in other stores, and attempting at completing the SKU-size plans of each store to the greatest feasible extent through an extension to our approach.

3 Methodology and Implementation

The whole process consists of four main steps (Fig. 1) as follows:

1. Head office-driven *replenishment suggestion* and *proposal to the stores*. SKU/size/store allocated quantity calculated based on past sales, category seasonality,

Fig. 1 Process chart

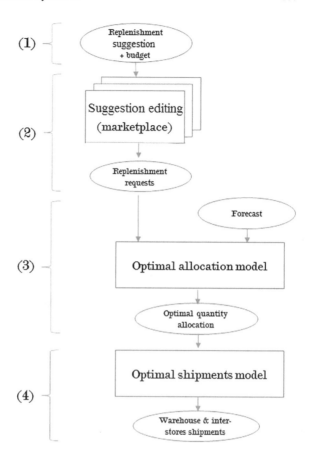

and size-level allocation. The output is a list of item/sizes to order and release. This list tries to approximate the store manager's expected contribution to the forecast in case he does not edit it. Moreover, each store is assigned a budget based on the current balance between current stock and expected (potential) future sales. Suggested orders consumes the budget, releases free it up.

2. Store-driven *internal marketplace*. During a given time window, store managers can modify the proposed non-mandatory suggestions freely, while always operating within the budget constraints (e.g., quantities of new items that can be ordered are capped by the budget, but additional budget can be freed up by releasing items currently in stock).

3. Head office-driven *optimal allocation*. The expected profit is maximized through the optimization of stock movements. The problem formulation contemplates expected demand, warehouse availability, logistic costs and stock availability constraints. The output of the optimization includes both deliveries from the central warehouses as well as direct transhipments across stores.

4. *Logistics*. Finally, orders and releases are optimally matched (head office-driven, minimizing shipments) and the parcels are shipped.

3.1 Notation

The notation used hereafter is as follows:

- J is the set of stores, and j is a specific store;
- I is the set of items, i is a specific item, and I_C is the set of items of a specific category C;
- S is the set of sizes, and s is a specific size; d is a specific sizing system (e.g., EU or US system), and S_d is the subset of sizes of sizing system d;
- w is a generic week, and $\tilde{w}$ is the current week; the set of weeks is denoted as W;
- P_{ij}^w is the full price for item i in store j at week w, $\tilde{P}_{ij}^w$ is the pocket price paid;
- E_{ijs}^w is the quantity in stock for item i and size s at the store j at the end of week w;
- H_{is}^w is the quantity available in the warehouse for item i and size s at the end of week w;
- n_f is the number of future weeks for which the replenishment is planned, and n_p is the number of past weeks to use as a reference for a first replenishment suggestion;
- Y_{ijs}^w is the quantity of item i and size s sold in store j at week w;
- $\tilde{Y}_{ijs}^*$ is the sales forecast (n_f weeks) for item i and size s in store j at week w;
- the omission of indexes i or j or s indicates a sum (e.g., Y_j^w is the quantity sold in store j at week w, all sizes and items).

Available data includes point of sales (POS) transactions and their related master data: item categories, unit costs, store coordinates, name and type, daily store traffic, stock levels in the central warehouse and at each store.

Historical data is netted of returns and sales aggregated at different levels (e.g., category, store, and SKU level), calculating total quantity sold, average price, average markdown, total traffic, and total tickets, at each level of aggregation.

3.1.1 Replenishment Suggestion

The replenishment suggestion together with the budget serves as a starting point for store managers and also provides an automated contribution in case the store managers decide not to modify the suggestions (or are not able to). Hence, the rationale is to imitate the reasoning that could lead an average (imaginary) store manager to a request of items to be replenished. A roughly estimated sale potential by SKU/store is calculated multiplying the past weeks' sales by a future seasonality coefficient.

The number of future weeks for which the replenishment is planned n_f can vary in time and depend on the retailer's supply chain policy. It can be low (even 1 or 2 weeks) for fast rotation retailers, higher (e.g., 6–8 weeks) for low rotation or high-end retailers. The number of past weeks to use as a reference for a first replenishment suggestion n_p has to be fixed in order to account for recent item mix and seasonality and to guarantee at the same time a reliable baseline for future forecast.

3.1.2 SKU-Store-Level Allocation

At the current week $\tilde{w}$, the total sales of item i in store j for the last n_p past weeks is given as

$$Y_{ij}^* = \sum_{w=\tilde{w}-4}^{\tilde{w}-1} \sum_{s \in S} Y_{ijs}^w.$$

The seasonality coefficient is calculated using weekly seasonalities z_C^w for each item category C (see Appendix A for the details). The estimated sales' potential at the current week $\tilde{w}$ $\tilde{Y}_{ij}^*$ for the next $n_f - 1$ following weeks is then calculated as

$$\tilde{Y}_{ij}^* = Y_{ij}^* \frac{\sum_{k=0}^{n_f} z_C^{\tilde{w}+k}}{\sum_{k=1}^{n_p} z_C^{\tilde{w}-k}}.$$

3.1.3 Size-Level Allocation

As we need size-level quantities, $\tilde{Y}_{ij}^*$ is allocated by size following the method described in Sect. 3.1.3, yielding $\tilde{Y}_{ijs}^*$. Item size-level allocation is typically performed based on the assumption of historical consistency and homogeneous store distribution of sales by size (Thomassey 2005, p. 87). The proposed systems aim to realize a more flexible allocation of sizes, adapted to the store characteristics and to the distribution of sales by size for a particular item. This aspect is often neglected, due to the complexity introduced by multiple sizing systems and to the high fragmentation of the information.

In order to distribute the SKU/store quantity forecast by size, we compute the relative frequency of a certain size s for a certain item and for a certain store. The relative frequency of sales of size s for item i is denoted by $p_i(s)$ and is defined as total sales of item i and size s versus the total sales of item i in the whole store network:

$$p_i(s) = \frac{\sum_{w \in W} \sum_{j \in J} Y_{ijs}^w}{\sum_{w \in W} \sum_{j \in J} \sum_{s \in S} Y_{ijs}^w} \quad \forall s \in S.$$

In order to define the relative frequency of size s for each store p_j, we first need to consider the sizing system of size s, denoted by d_s. Then, we compute for each store j the ratio between sales of size s and total sales of all sizes of the same sizing system d_s:

$$p_j(s) = \frac{\sum_{w \in W} \sum_{i \in I} Y_{ijs}^w}{\sum_{w \in W} \sum_{i \in I} \sum_{s \in S_{d_s}} Y_{ijs}^w} \quad \forall s \in S.$$

Notice that using the frequency across sizes of sales of item i in store j would not be accurate, as volumes can be low.

Figure 2 shows example distribution of sizes for different stores and items.

Both factors are equally weighted, in order to take into account both the item and the store characteristics separately:

$$p_{ij} = 0.5 \cdot p_i + 0.5 p_j. \tag{1}$$

To forecast future sales $\tilde{Y}_{ijs}^*$ for each item and size, we generate a random sample according to the distribution p_{ij} of sizes, to match the total forecast $\tilde{Y}_{ij}^*$.

The initial proposal dramatically reduces the workload for store managers, thanks to automated allocations to sizes and balancing of demand across items.

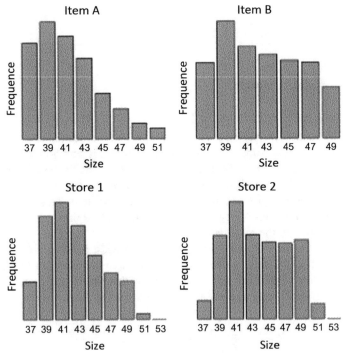

Fig. 2 Frequencies of pieces sold by size, for two items (all stores) and for two stores (all items)

3.1.4 Proposal to the Stores

The store-level budget is a simple measure of the gap between the current stock levels and the future expected sales potential; it gives store managers a transparent, non-monetary incentive to release non-performing stock so they can order new items; it also increases empowerment and direct ownership of stock allocation decisions: all stock received will always have been requested, even if not all the requests can be fulfilled due to system constraints.

For each store j, the budget b_j at current week $\tilde{w}$ is calculated as the sales expected overall potential for future n_f weeks plus a safety extra coverage of 20%, minus the total potential value of current stock:

$$b_j = (1.2) \cdot \sum_i \tilde{Y}_{ij}^* P_{ij}^{\tilde{w}} - \sum_i \left(\sum_{s \in S_i} E_{ijs}^{\tilde{w}-1} \right) P_{ij}^{\tilde{w}},$$

where $P_{ij}^{\tilde{w}}$ is the selling price of item i in store j, and $E_{ijs}^{\tilde{w}}$ is the current stock of item i, size s in store j. Then, $\tilde{b}_j = b_j - \frac{\sum_j b_j}{|J|}$, calibrating all budgets so that the median across stores is zero.

If the budget is positive, then the store is *relatively understocked* and therefore needs additional stock. Otherwise, the store is *relatively overstocked* and should therefore release stock; Table 1 shows a simple illustrative example.

	Item 1	Item 2	Item 3
Selling price $P_{ij}^{\tilde{w}}$	100	200	150
Pieces forecast $\tilde{Y}_{ij}^*$	5	3	3
Stock E_{ij}^{w}	2	4	2

The budget is a measure of the store stocking level. However, it does not indicate how the stock level is distributed over the SKUs and sizes. Thus, we call Δ_{ijs} the difference between stock and potential sales at store/SKU/size level:

$$\Delta_{ijs} = E_{ijs}^{\tilde{w}-1} - \tilde{Y}_{ijs}^*.$$

Table 1 Example of store budget definition for store j with three items

Expected revenues = $5 \times 100 + 3 \times 200 + 3 \times 150 = 1550$
Stock potential value = $2 \times 100 + 4 \times 200 + 2 \times 150 = 1300$
Extra coverage = $20\% \times 1550 = 310$
Budget = $1550 - 1300 + 310 = +560$

Table 2 Recommended actions based on quartiles

$\Delta_{ijs}\%$ quartile	$\Delta_{ijs} > 0$ (high stock)	$\Delta_{ijs} < 0$ (low stock)
1 or 2 (low)	No action	No action
3 (medium)	Recommended release	Replenishment
4 (high)	Mandatory release	Urgent replenishment

Table 3 Example recommended actions

Store	Item	Stock	Potential	Delta	Ratio	Quartile	Action
Store 1	Item 1	10	1	9	9/11 = 0.82	4	Mandatory release
Store 1	Item 2	1	10	−9	9/11 = 0.82	4	Urgent replenishment
Store 2	Item 1	3	1	2	2/4 = 0.5	3	Recommended release
Store 2	Item 2	6	5	1	1/11 = 0.09	1	No action
Store 3	Item 1	7	8	−1	1/15 = 0.07	1	No action
Store 3	Item 2	2	5	−3	3/7 = 0.43	2	Replenishment

We rank orders and releases based on their relative priority, through the ratio:

$$\Delta_{ijs}^{\%} = \frac{|\Delta_{ijs}|}{E_{ijs}^{\tilde{w}-1} + \tilde{Y}_{ijs}^{*}}.$$

The recommendation for each item is based on its Δ_{ijs} and $\Delta_{ijs}^{\%}$ quartiles for each store:

- if $\Delta\%$ is low, stocking of item i, size j in store s can be considered sufficient and no action is needed
- is $\Delta\%$ is very high, the stock is not balanced with potential sales; a urgent replenishment (if $\Delta < 0$) or forcing the release of the item (if $\Delta > 0$) is needed
- for mid-levels of $\Delta\%$, stock and potential sales are quite unbalanced; a replenishment (if $\Delta < 0$) or a release of the item (if $\Delta > 0$) is suggested

A summary of recommended actions is given in Table 2, while an example of item-level analysis is in Table 3.

3.2 Extra Features of the Proposal

If all releases and replenishment would be actually done, the stock level would be balanced across the whole store network. However, after considering recommended actions, some further adjustments to the *proposal* can be made, in order to account for additional business requests:

- store demands can be capped at the total quantity available from warehouse inventory and other stores (releases)
- items introduced for the first time in the last (e.g., two) weeks can be excluded from release recommendations
- a minimum quantity (e.g., one piece) for each item/size can be left in each store
- orders for sizes that are currently out of stock can be prioritized

The proposed system can be also used to facilitate logistic and merchandising activities by accounting for orders and requests that are due to supplementary criteria, not directly related to demand and stock level. For example, it is possible to use the system to reduce the fragmentation of the stock, i.e., to reduce the number of stores where stock-out occurs for most of the sizes of a certain SKU.

In order to take into account this, fragmentation has to be denined measured at the store/SKU level. The definition takes into account the number of different sizes in which the SKU is sold. If the percentage of sizes of SKU i in store s that are not available is greater than a percentage threshold t_{f1}, item i is defined *fragmented at the store level* in store s. We define the variable φ_{is} as one if i is fragmented at the store level in s, otherwise zero. Then, for all SKUs i, we define:

$$F_i = \{s \in S | \exists (i, s), E_{ijs}^{\tilde{w}} > 0, \varphi_{is} = 1\}$$

$$T_i = \{s \in S | \exists (i, s), E_{ijs}^{\tilde{w}} > 0\}$$

Finally, we compute for each SKU the ratio between the number of stores in which i is fragmented and the number of stores in which i is available, that is

$$R_i = \frac{|F_i|}{|T_i|}$$

And we define the SKU i *globally fragmented* if R_i is higher than a threshold t_{f2} (e.g., 40%). It is possible to modify store requests in order to reduce the fragmentation in the whole retail network. One possibility is that stores with highest potential order one piece of each size of globally fragmented SKUs and stores with lowest potential release the same SKUs.

3.3 Internal Marketplace

Store managers can edit the initial proposal as they see fit, based on their relevant local knowledge, except for mandatory instructions. The stores can request items and sizes outside the initial proposal, and even if not yet sent to the store; they can also change the quantity of pieces ordered.

However, they must respect their budget constraint. This forces the store manager to accurately choose orders and facilitates the release of unsold items, that can be

Table 4 Extract of proposal and edited proposal for a store (example)

SKU/Size	Pcs sold	Stock	Sugg action	Budget (sugg)	Req action	Budget (req)
Item1/37	0	2	Rel 1pc	+129	Rel 2pc	+258
Item1/41	2	3	Rel 1pc	+129	Rel 2pcs	+258
Item1/43	0	2	MRel 1pc	+129	MRel 1pc	+129
Item1/47	0	2	MRel 1pc	+129	MRel 1pc	+129
Item1/49	1	1	Order 1pc	−129	Order 1pc	−129
Item1/51	0	0	–	–	Order 1pc	−129
Item1/53	0	0	–	–	Order 1pc	−129
Item2/37	2	2	Rel 1pc	+99	Rel 1pc	+99
Item2/39	2	2	Rel 1pc	+99	Rel 1pc	+99

shipped to stores in which they are considered more attractive. This way, the system gives store managers the possibility to decide what to offer, but also guarantees that the allocation is fair and transparent.

At the end of the assigned time window to make changes, the modified proposals are collected. Table 4 shows an extract of an edited proposal example. For each SKU/size, the store manager can see the *suggested action* (Orders/Releases/Mandatory Releases) and their contribution to the budget. Then, the store manager edits the suggestion writing *requested actions*. In the example, the store manager has made new orders but also increased the releases so to match the budget constraint.

3.3.1 Popularity Index

Additionally to direct monetary benefits, the new system provided the central merchandising team with a frequent "survey-like" review of customers' preferences, based on the requests of store managers (as a proxy). This can help to predict slow and fast movers, both locally as well as globally.

In order to measure and control store managers' preferences, we defined a *popularity index* that takes into account the intensity of orders or releases for each SKU and their inventory level. For each item i, we define

$$\phi_i^w = \min\left(1, \frac{\sum_{s,j} R_{ijs}^{\tilde{w}}}{\sum_{s,j} E_{ijs}^w}\right) \tag{2}$$

The numerator in (2) is the net total request for item i, on all stores and sizes; the denominator is the total inventory in the stores involved in the replenishment program. The resulting quantity is bounded, i.e., $\phi_i^w \in [-1, 1]$. Ranking all the SKUs of the brand at a fixed week w, an S-shape curve in Fig. 3 is found.

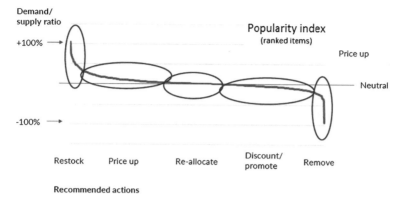

Fig. 3 Ranked popularity index of SKUs of the brand in the test and recommended actions

It is then possible to classify the SKUs in groups according to their popularity. The management can then consider strategic actions. For example, SKUs with highest popularity can be urgently restocked from central warehouse or other sources; other SKUs with positive index can be recommended for a price increase if they will not be restocked. On the other side, SKUs with negative index can be recommended for promotional activities, even during the season. If $\phi_i^w \sim -1$ s, all stores would prefer to replace the SKU with others, so the retailer can consider to remove it from the assortment.

The central part of the S-curve in Fig. 3 represents the group of SKUs for which $\sum_{s,j} R_{ijs}$ is almost null. Those items are likely to be the most profitable to reallocate, due to opposite requests made by different stores.

3.4 Optimal Allocation

Once proposals of every store managers are compiled, the allocation needs to be optimized. We base the allocation process on three calculation steps:

1. Demand forecast for week $\tilde{w}$ at item/size level for each store (adapted from Correa (2007))
2. Expected sales as a function of stock constraints, for each item in each store
3. Optimal allocation, maximizing the expected future profit

At week $\tilde{w}$, let $R_{ijs}^{\tilde{w}}$ be the number of pieces ordered by the store managers, if positive, or released, if negative. Moreover, let $x_{ijs}^{\tilde{w}}$ be the quantity (item i, size s, and store j) to be allocated. If $x_{ijs}^{\tilde{w}} > 0$, the store will receive the corresponding articles; otherwise, if $x_{ijs}^{\tilde{w}} < 0$ the store is expected to release the goods and ship them to other stores.

3.4.1 Demand Forecast

It is possible to consider the number of sold pieces of a certain SKU/size/store as the count of occurrences of a random event that happens at a certain rate. In this framework, we can model sales as a Poisson process. Let $N_{ijs}^w(t)$ be the Poisson process that counts the number of items sold in the time interval $(w, w + t]$ with intensity λ_{ijs}^w. The parameter λ_{ijs}^w is estimated as the mean number of articles expected to be sold in week w, calculated as follows.

For a given category C, let $\tilde{Y}_C^{\tilde{w}}$ be the sales forecast (aggregated by sizes, stores, and items of the category C) of week $\tilde{w}$ calculated as described in Appendix A. To derive the sales forecast for a single item/size in a given store, the allocation is computed:

- by store, according to the ratio between past store sales for items in C and the total past sales for items in C across all stores, during the previous year
- by item, according to the proportion of store managers' positive requests for the item versus the total positive requests for items in C
- by size, using the relative frequency for size s of item i already derived in Sect. 3.1.3

So, we obtain:

$$\tilde{Y}_{ijs}^{\tilde{w}} = \tilde{Y}_C^{\tilde{w}} \cdot \frac{\sum_{w \in W^*} \sum_{i \in C} \sum_{s \in S} Y_{ijs}^w}{\sum_{j \in J} \sum_{w \in W^*} \sum_{i \in C} \sum_{s \in S} Y_{ijs}^w} \cdot \frac{\sum_{s \in S} \max(R_{ijs}^{\tilde{w}}, 0)}{\sum_{i \in C} \sum_{s \in S} \max(R_{ijs}^{\tilde{w}}, 0)} \cdot p_{ij}(s), \tag{3}$$

where R_{ijs}^w are the store managers' requests, W^* is the set of weeks corresponding to the previous year, and $p_{ij}(s)$ is the size frequency described in Sect. 3.1.3. Finally, the estimated mean expected sales are based on a combination of forecast and store requests, as follows:

$$\lambda_{ijs}^{\tilde{w}} = \alpha \cdot \left[E_{ijs}^{\tilde{w}-1} + R_{ijs}^{\tilde{w}} \right] + \beta \cdot \tilde{Y}_{ijs}^w, \tag{4}$$

where α and β are two coefficients such that $\alpha + \beta = 1$. The weights assigned to the manager's experience and the statistical forecasting can be updated based on the relative accuracy observed. The better the measured performance of store managers at predicting future sales, the larger the α.

3.4.2 Expected Sales as a Function of Stock Constraints

The demand forecast so far does not account for stock constraints; however, the maximum potential sales depend on the store stock. Thus, the simple Poisson process is not enough to model sales. Assuming mutual independence across the sales of different sizes and items, and that the average intensity of the process is equal to the demand forecast $\lambda_{ijs}^{\tilde{w}}$ computed in Sect. 3.4.1, it is possible to show (see Sirovich et al. (2018)) that the expected sales in a time period of T weeks until stock-out equals

$$\sum_{k=1}^{E_{ijs}^{\tilde{w}-1}+x_{ijs}^{\tilde{w}}} \mathbb{P}(N_{ijs}^{\tilde{w}}(T) \geq k),$$

which yields the expected sales as a function of forecast demand (see Sect. 3.4.1) and new stock allocated.

3.4.3 Logistics

The total expected income can be expressed as a function of the items' allocation among stores $x_{ijs}^{\tilde{w}}$.

First, the expected income for item i, size s, across all stores, is calculated by multiplying the expected sales and the price $P_{ij}^{\tilde{w}}$ of the item in each store.

$$\sum_{j \in J} P_{ij}^{\tilde{w}} \sum_{k=1}^{E_{ijs}^{\tilde{w}-1}+x_{ijs}^{\tilde{w}}} \mathbb{P}(N_{ijs}^{\tilde{w}}(T) \geq k).$$

Then, the potential profit from items remaining in the warehouse after the shipments is:

$$K \cdot \left(H_{is}^{\tilde{w}-1} - \sum_{j \in J} x_{ijs}^{\tilde{w}} \right).$$

These items can still be used for future shipments, so the value K corresponds to their expected future income. K is larger at the beginning of the season and of the item's life cycle, then progressively smaller as the remaining shelf time left gets shorter.

Finally, the unknown $x_{ijs}^{\tilde{w}}$ are calculated to maximize, for each i and s, the total expected net income:

$$\sum_{j \in J} P_{ij}^{\tilde{w}} \cdot \left(\sum_{k=1}^{E_{ijs}^{\tilde{w}-1}+x_{ijs}^{\tilde{w}}} \mathbb{P}\left(N_{ijs}^{\tilde{w}}(T) \geq k\right) \right) + K \cdot \left(H_{is}^{\tilde{w}-1} - \sum_{j \in J} x_{ijs}^{\tilde{w}} \right), \quad (5)$$

where T is the time horizon of the procedure. The optimization is subject to the following constraints:

$$(a) \quad x_{ijs}^{\tilde{w}} \leq \max\left(R_{ijs}^{\tilde{w}}, 0 \right)$$

$$(b) \quad x_{ijs}^{\tilde{w}} \geq \min\left(R_{ijs}^{\tilde{w}}, 0 \right)$$

$$(c) \quad 0 < \sum_{j \in J} x_{ijs}^{\tilde{w}} < H_{is}^{\tilde{w}-1}, \quad (6)$$

Table 5 Summary of optimal allocation output (extract example)

Store	Budget [EUR]	Sugg O	Sugg R	Req O	Req R	Fulf O	Fulf R	% proc
WH	−7653	0	0	0	0	0	0	0
1	−7386	155	198	155	198	102	175	77
2	−2827	0	0	30	230	11	10	13
3	−2069	54	59	70	64	43	49	70
4	43095	356	0	383	22	219	0	56
5	26149	229	0	240	12	151	0	61
6	13037	184	62	203	134	95	32	41
7	−30356	0	202	0	74	0	61	84
8	−3762	0	0	1	0	0	0	70
9	−2916	0	0	52	144	23	2	19

where (a) and (b) ensure that the store requests are not exceeded (if $R_{ijs}^{\tilde{w}} > 0$) and that the maximum deliveries from the store do not exceed the released quantity (if $R_{ijs}^{\tilde{w}} < 0$). The last constraint (c) ensures that no items are sent back to the warehouse from the stores, and that (trivially) every piece is available from the stores or the central warehouse.

The optimization problem is an integer programming formulation, solved with the GENOUD algorithm (Mebane and Sekhon 2011). Example summary of optimal allocation output is given in Table 5, showing Orders and Releases for each store and for the warehouse (WH). The initial proposal (suggested), the store-reviewed proposal (requested), and the final output allocation (fulfilled) are shown. In this example, over 50% of the store managers' requests are satisfied.

Notice that the store budget b_j directly influences the store requests, but is not directly involved in equation, where individual stock levels by SKU and store are involved. The optimization process attempts to rebalance stock levels only if profitable extra shipment is possible. Thus, after the optimization the stock levels across the stores will be more balanced, but still not exactly balanced, avoiding non-profitable extra costs.

4 Pilot Study and Results

Our case study company, an Italian retailer, is a leader in the Italian premium curvy fashion retail, with approximately 100 stores and an average item selling price of 150 EUR.

The retail footprint of the retailer in analysis spans across ∼100 stores in Italy, collectively selling ∼400 k pieces a year. Two collections alternate every year, summer and winter, each consisting of ∼1300 SKUs grouped in 20 categories; as is typical

Fig. 4 Boxplot of the ratio
of the suggestion modified,
for all the test stores at week
8 of the test

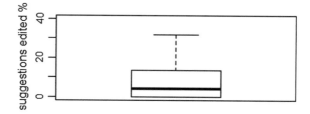

of the luxury segment, stores carry a low inventory of each item, but guaranteeing availability of all sizes in the item size plan at each store. Sales are highly seasonal. The company switched to a fully automated system managed centrally, that assumed a homogenous demand distribution across stores (with rare exceptions for special categories such as beachwear), with a simple scaling based on the overall level of sales for each store (rounded to the nearest integer).

This new process has been subject to a rigorous pilot experiment during the Spring-Summer 2016 season to estimate impact on sales and revenues, but also to collect feedback from the store managers.

We set up a classical test-control study: the retailer's management selected test and control stores based on their experience, as a representative sample of the store universe in terms of geographical location, sales volume, and historical seasonality. Test and control groups had, respectively, 18 and 46 stores, and the entire product range was included in the study.

During the 13-week test phase, the coefficients in (1) have been set to $\alpha = \beta = 1/2$ as it is the natural initial choice when the store manager's predictive skill is not evaluable.

The test group used the new replenishment process, while the control group kept the automatic headquarter-driven process. The number of future weeks for replenishment n_f was set to 8. Moreover, the number of past weeks n_p to use as a reference for the replenishment suggestion and budget computation was set to 4.

The stores actively took part to the test modifying the first suggestions. The average participation ratio (in terms of ratio of edited suggestions over total suggestions) was 72%. As shown in Fig. 4, store managers edited from 0 to 33% of total pieces in the proposal, and the median of the percentage of suggestion edited is 4%.

4.1 Test Impact Evaluation

Indeed, we compared test and control stores in terms of operational performance of allocation process. We used 4 performance metrics: 3 overstock measures (stock to the stores not to exceed demand) and 1 of understock (shipments sufficient to cover consumers' demand), explained in the following.

Shipment Success Ratio (SSR): the proportion of delivered items (positive replenishment) actually sold after delivery. For each positively delivered item, the ratio between gained and sold articles is calculated. As we want to measure the shipped items actually sold, the sales are capped to the total number of delivered items. This quantity is then cumulated over time, since items can be sold for few weeks after delivery. So finally, at week $\tilde{w}$ the SSR is given by

$$SSR_{\tilde{w}} = \frac{\sum_{w=1}^{\tilde{w}} shipments_w}{\sum_{w=1}^{\tilde{w}} \min\left(sales_w, shipments_w\right)}.$$

Sales to Shipment Ratio (StSR): the proportion of total sales over the number of delivered pieces. The StSR at week $\tilde{w}$ is then

$$StSR_{\tilde{w}} = \frac{\sum_{w=1}^{\tilde{w}} sales_w}{\sum_{w=1}^{\tilde{w}} shipments_w}.$$

If the number of sold items is larger than the number of injected items, this metric is larger than one and means that the stores have high stock levels, that should be monitored. Every unsold stock item is a potential sale lost.

Stock Velocity Ratio (SVR): the ratio between total sales and potential sales. The latter are evaluated as residual stock plus cumulated sales. The SVR at week $\tilde{w}$ is given by

$$SVR_{\tilde{w}} = \frac{\sum_{w=1}^{\tilde{w}} Sales_w}{Stock_{\tilde{w}} + \sum_{w=1}^{\tilde{w}} Sales_w}.$$

To complete the picture, we need a metric to evaluate wether the stores are better fulfilling the clients' expected demand. We call it *Demand Cover Ratio (DCR)*: the ratio between sold items and expected demand, calculated as in Caro and Gallien (2010). At each week w, if sales are positive and the number of stock-out days is smaller than 7, the expected demand is:

$$demand_w = sales_w \cdot \left(\frac{7}{7 - \text{stock-out days}_w}\right).$$

Otherwise, expected demand is set to the most recent nonnegative demand (or zero). The rationale is that potential demand is equal to actual sales plus missed sales due to stock-outs. So finally, DCR at week $\tilde{w}$ is given as

$$DCR_{\tilde{w}} = \frac{\sum_{w=1}^{\tilde{w}} sales_w}{\sum_{w=1}^{\tilde{w}} demand_w}.$$

The metrics have been calculated for test and control groups, and results are illustrated in Table 6.

Table 6 Monitoring metrics: test (T) versus control (C) KPIs

Week	(T) SSR %	(C) SSR %	(T) StSR%	(C) StSR%	(T) SVR%	(C) SVR%	(T) DCR%	(C) DCR%
1	12	12	166	161	6	5	58	57
2	19	17	111	112	12	9	59	56
3	22	19	86	76	17	13	60	58
4	30	24	93	76	24	19	60	60
5	31	27	88	68	27	21	61	62
6	35	30	90	73	28	23	61	61
7	40	33	91	74	31	26	61	61
8	43	36	100	81	34	28	60	61
9	43	39	92	89	36	31	60	60
10	47	42	99	94	38	33	60	60
11	51	46	106	101	41	36	60	59
12	53	50	112	110	44	39	60	59
13	58	53	132	123	49	43	61	60

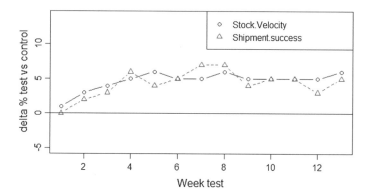

Fig. 5 Weekly difference between test and control indexes SSR and SVR

Figure 5 shows how SSR and SVR increase for Test stores since the beginning of the test, and that this increase remains stable along the prosecution of the test.

At the same time, DCR Fig. 5 shows how DCR of Test and Control stores remained similar (i.e., extra sales coming from shipments do not lead to higher occurrence of stock-out in the shipping store). StSR sharply increased especially in the first weeks.

On average, the pilot demonstrated the following:

- up to 7% increase in SSR, and up to 19% increase in StSR, a significant improvement in the relevance of shipments compared to the past, resulting in more sales;
- up to 6% increase in SVR; greater stock rotation leading to a reduction in unsold stock;

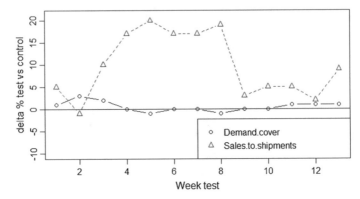

Fig. 6 Weekly difference between test and control indexes StSR and DCR

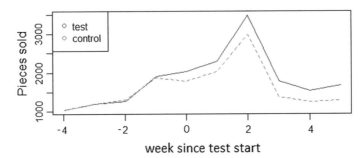

Fig. 7 Impact on sales: test versus scaled control

- <1% greater DCR, suggesting further opportunity to improve the coverage of latent demand and further reduce the risk of stock-outs for items with additional potential demand (Fig. 6).

These results, in monetary terms, translated into an estimated 280 k EUR margin impact during this pilot. The positive impact on sales of the test stores is shown in Fig. 7, where control sales were scaled to test group sales before the test. The estimated impact was a significant +16% revenues, worth 280 k EUR margin impact just during this pilot (over three months and 18 stores). We expect a margin of 1M/month over the entire network for the entire year, considering a 40% lower expected impact after the end of the test phase.

As non-monetary benefits, store managers' positive feedback, sense of ownership and empowerment, resulted in a boost of staff morale. The store managers became more motivated to sell the items they had specifically requested.

As it was mentioned before, the requests of the store managers provide the retailer central office a survey on the quality of each item. Figure 8 shows a correlation between SKU-level popularity index and sell through for each SKU, meaning that store managers' orders have a good predictive power.

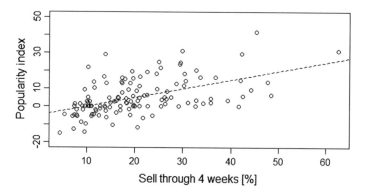

Fig. 8 SKU-level popularity index and sell through for each SKU. The chart shows a correlation between the two, meaning that store managers' orders have a good predictive power

5 Conclusions

The purpose of this work was to show the implementation and results of an original intelligent system able to improve the replenishment process of fashion retail companies, in order to limit missing sales due to stock-out, enrich, and adapt the assortment of the single stores to local preferences. We combined a traditional forecast-driven approach to the requests of store managers, in order to achieve a greater level of store customization and ultimately greater local relevance to the specific demand of each store.

Each store manager receives an initial proposal they can modify, within certain budget constraints, creating a non-monetary incentive to accurately assess her expectations on the relative expected performance of the current stock, and decide what to release and what to request from other stores, thus creating an internal marketplace.

All requests are matched to the available inventory across all stores, and deliveries are optimized to minimize the number of parcels (and logistic costs) while maximizing the expected profit based on a greater expected likelihood of sale.

We successfully tested the system in a 13-week pilot. Test stores outperformed the control group on all the metrics and also increased sales relative to control.

However, this new process is open to many further improvements. Transhipments across stores could be optimized in order to further reduce logistic costs and store managers' workload. For example, a minimum shipment value can be imposed, or a cap to the maximum number of parcels that can be delivered by a store might be enforced.

Furthermore, the pilot store managers personally recommended some further additional improvements, mainly to reduce the size-level fragmentation and probability of stock-outs, for example, by penalizing releases that would lead to a stock-out, or by modifying the system so it would re-compact SKUs across their entire size plan.

Acknowledgements The authors gratefully acknowledge Evo Pricing for supporting this research.

Appendix: Forecasting Model

Category sales $\tilde{Y}_C^{\tilde{w}}$ (aggregate: all sizes, all stores, and items in C, with one-year time horizon) are forecasted through multivariate regression on seasonality $\bar{z}$, average full price p, average markdown m, and units per tickets ratio u. The forecasting procedure consists in these steps:

1. historical data is aggregated at the category, year, and week level
2. for every category C and week of the year w, weekly seasonality index $\bar{z}_C^w$ is computed
3. a linear model $\tilde{Y} = f(z, p, m, u)$ is fitted for every category
4. an estimation of the KPI p, m, and u is given for the following 52 weeks
5. one-year horizon sales forecast is computed $\tilde{Y}_C^w$ for each category and week by using the estimated models

In the following, details on seasonality index and estimation of the regressors for the future year are provided.

A. Category Seasonality

From the weekly aggregated sales, for each category, we remove the effect of markdowns by estimating the model

$$Y_C^w = \alpha_C + \beta_C m d_C^w.$$

Then for each week, we calculate normalized sales

$$\bar{Y}_C^w = Y_C^w - \beta_C m d_C^w.$$

We then average normalized sales $\bar{Y}_C^w$ over the years, obtaining $\bar{\bar{Y}}_C^w$, and we smooth this using the moving average of three weeks:

$$z_C^w = \frac{\bar{\bar{Y}}_C^{w+1} + 3\bar{\bar{Y}}_C^w + \bar{\bar{Y}}_C^{w-1}}{5}.$$

Finally, we normalize seasonality (every category sums to 52)

$$\bar{\bar{z}}_C^w = z_C^w \times \frac{52}{\sum_w z_C^w}.$$

B. Prediction of Business Indicators

Future average values of full price p, markdown m, and units per tickets ratio u are estimated by moving averages. As an example, starting from category level average full price for a given week and year, p_C^w, we average price over the years obtaining $\bar{p}_C^w$. Then, we smooth this using the moving average of three weeks:

$$\bar{\bar{p}}_C^w = \frac{\bar{p}_C^{w+1} + 3\bar{p}_C^w + \bar{p}_C^{w-1}}{5}.$$

$\bar{\bar{m}}$ and $\bar{\bar{u}}$ are then computed in the same way and then used with in (5) as regressors together with $\bar{p}$ and $\bar{\bar{z}}$. It is also possible to manually change these values according to future management strategy.

References

Agrawal N, Smith SA (2013) Optimal inventory management for a retail chain with diverse store demands. Eur J Oper Res 225(3):393–403

Blattberg RC, Hoch JS (1990) Database models and managerial intuition: 50% model + 50% manager. Manag Sci 36(8):887–899

Cachon GP, Lariviere MA (1999) An equilibrium analysis of linear, proportional and uniform allocation of scarce capacity. IIE Trans 31(9):835–849

Caro F, Gallien J (2010) Inventory management of a fast-fashion retail network. Oper Res 58(2):257–273

Correa J (2007) Optimization of a fast-response distribution network. M.S. thesis, LFM, MIT, Cambridge, MA

Fisher M, Rajaram K (2000) Accurate retail testing of fashion merchandise: methodology and application. Mark Sci 19(3):266–278

Furuhata M, Zhang D. Capacity allocation with competitive retailers

Gallien J, Mersereau AJ, Garro A, Mora AD, Vidal MN (2015) Initial shipment decisions for new products at Zara. Oper Res 63(2):269–286

Mebane WR Jr, Sekhon JS (2011) Genetic optimization using derivatives: the rgenoud package for R. J Statist Softw 42(11):1–26

Sirovich, Marocco, Craparotta. A woman's touch in fashion forecasting: combining analytics & expert judgement, in preparation

Thomassey S (2010) Sales forecasts in clothing industry: the key success factor of the supply chain management. Int J Prod Econ 128(2):470–483

Thomassey S, Happiette M, Castelain JM (2005) A global forecasting support system adapted to textile distribution. Int J Prod Econ 96(1):81–95

Van Donselaar KH, Gaur V, Van Woensel T, Broekmeulen RA, Fransoo JC (2010) Ordering behavior in retail stores and implications for automated replenishment. Manag Sci 56(5):766–784

Blockchain-Based Secured Traceability System for Textile and Clothing Supply Chain

Tarun Kumar Agrawal, Ajay Sharma and Vijay Kumar

Abstract Blockchain has emerged as a prominent and reliable solution that can enable and ensure secure information sharing over wide area networks. In an era of digitalisation, blockchain technology is finding wide applications in multiple fields including implementing traceability in the supply chain. In this direction, this chapter explores its potential application in implementing a blockchain-based traceability system for textile and clothing (T&C) supply chain. It examines the necessity and concept of a traceability system, followed by enlisting advantages of blockchain technology for implementing traceability. Further, a case-based example has been used to explain blockchain application in implementing traceability in T&C supply chain. Finally, it mentions the challenges and limitations of such blockchain-based traceability system that can be addressed through further research.

1 Introduction

Information asymmetry, visibility and security are some of the key concerns for multi-tier supply chains (Caridi et al. 2010; Bhargava et al. 2013; Tachizawa and Wong 2014). The supply chain actors often it difficult to recognise and access

---

The original version of this chapter was revised: Incorrect co-author affiliation has been corrected. The erratum to this chapter is available at https://doi.org/10.1007/978-981-13-0080-6_15

T. K. Agrawal
The Swedish School of Textiles, University of Boras, 50190 Boras, Sweden

T. K. Agrawal · V. Kumar (✉)
ENSAIT, GEMTEX – Laboratoire de Génie et Matériaux Textiles, 59000 Lille, France
e-mail: vijay.kumar@ensait.fr; vijay856@gmail.com

T. K. Agrawal
Univ Lille Nord de France, 59000 Lille, France

T. K. Agrawal
College of Textile and Clothing Engineering, Soochow University, Suzhou, China

A. Sharma
Independent Consultant, Roubaix, France

© Springer Nature Singapore Pte Ltd. 2018
S. Thomassey and X. Zeng (eds.), *Artificial Intelligence for Fashion Industry in the Big Data Era*, Springer Series in Fashion Business,
https://doi.org/10.1007/978-981-13-0080-6_10

all the involved supplier and sub-suppliers (Grimm et al. 2016). Moreover, with globalisation and supply chain diversification, these issues have further intensified in the past few decades.

Textile and clothing (T&C) supply chain is one such example that is suffering significantly with these challenges (De Brito et al. 2008; Kumar 2017). Well acknowledged as one of the major employers in developing countries, T&C industries have also been in the limelight from time-to-time for their unsustainable practices in the form of child labour, uneven wages, sweatshops, improper treatment/discharge of toxic waste, etc. (Kumar et al. 2017b). Growing market demands, shorter shelf life and quest for cheap product are some of the major reasons for adoption of such unethical production means (Caniato et al. 2012; Kumar et al. 2017a). Nonetheless, serious catastrophic events in recent past have accelerated the demands for more transparent T&C supply chain (Kumar et al. 2017c). Governments are under constant pressure from customers and social welfare organisations to take actions and implement mandatory information sharing and transparency rules to regulate T&C supply chains and control these practices.

On the other hand, opaque and unsecure supply chains pave the way for counterfeits that have always been an important concern especially for the T&C industry (Corbellini et al. 2006). These counterfeit products not only affect the fashion brand image and economy, but are also harmful for the consumer health due to their inferior quality (Ekwall 2009). Every year the customs and border security forces confiscate counterfeit products worth millions of dollars. However, their trade is practically uncontrollable due to the huge volumes and numerous weak links in the T&C supply chain. As per the report from the European Union Intellectual Property Office Observatory, due to counterfeit products, the T&C industry loses 9.7% of sales, 26.3 billion euro of revenue per year in the sector, 36,300 direct jobs and 8.1 billion euro of revenue by government (Wajsman et al. 2015)

Traceability, as mentioned in the *OECD report* (2017), is one of the mechanisms that can be adopted to counter these issues. In general, traceability is an essential tool that can be used to gain access to information related to all supply chain actors and activities. It is among the significantly researched and widely adopted mechanism in the food and pharmaceutical supply chain (Peres et al. 2007). However, despite numerous benefits, traceability is a comparatively less explored and adopted mechanism in T&C supply chain due to lack of dedicated mechanism that can cater to the challenges of T&C supply chain. In this context, the current chapter introduces traceability and its advantages and further proposes blockchain-based secured traceability systems that can be adopted to counter the above-stated issues.

2   Understanding T&C Supply Chain

T&C industry is one of the oldest manufacturing sectors growing out from the industrial revolution back in the eighteenth century (Kumar 2017). Presently, it is one of the major contributors in the world economy. The global market for textile mills valued $667.5 billion and for apparel valued $842.7 billion in 2015, which is

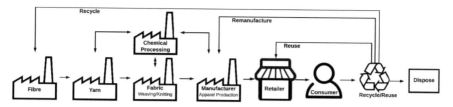

Fig. 1 A typical textile supply chain

expected to reach $842.6 billion and $1004.6 billion, respectively, by 2020–2021 (Marketline Report 2017). The T&C supply chain involves numerous processes, diverse raw materials and multiple intermediate products. Post multi-fibre agreement (MFA) abolishment, the T&C supply chain became more complex and distributed globally with huge geographical distance among the different stakeholders (Yang et al. 1997). Most of the manufacturing operations shifted to the developing countries like China, India and Bangladesh, whereas developed countries dominated the retail, due to population with comparatively high disposable income.

The T&C supply chain starts with the fibre cultivation in the form of natural fibre like cotton, wool and hem or chemical generation of the synthetic fibre like polyester, polyamide and acrylic (Doorey 2011). These fibres further pass to the spinning stage to formulate as yarns, which are then woven or knitted to form the sheet of fabric. These fabrics undergo several pre-treatment before they are shaped into 3D apparel or any other textile products. Depending on the application and functionality, these apparels or textile products are subjected to final finishing processes to make them fit for retailing. A typical T&C supply chain network is illustrated in Fig. 1.

Apart from its complex, diverse and opaque nature of supply chain, the T&C industry is also one of the most polluting industries. In this era of fast fashion, T&C industry is highly criticised for overproduction and overconsumption of textile products with least importance to recycling, remanufacturing and reusing the textile products. Most of the textile waste goes into landfills, causing hazardous effect on the soil.

3 Traceability

As per ISO 9000:2015, *Traceability is the ability to identify and trace the history, distribution, location, and application of products, parts, materials, and services*. A traceability system records and follows information trail of products, parts, materials and services come from suppliers and are processed and ultimately distributed as final products and services. Traceability supports the three pillars of sustainability, i.e. economic, environmental and societal (Kumar et al. 2017a). It facilitates effective information flow and sharing among different supply chain actors, thus helps in creating transparency, product data management, demand forecasting and logistics management (Agrawal et al. 2016). The following are some of the areas where traceability contributes.

(a) **Transparency**: Allowing on-demand and effective information sharing, traceability results in more transparent supply chain. A traceability system records and shares all the essential information from different supply chain actors. Thus, it helps in tracing the origin of a product and examining its actual social and environmental impact.

(b) **Quality Management**: With growing numbers of recall and low-quality products in T&C supply chain, an effective traceability system can help in quality monitoring purpose. The origin of defect and involved stakeholders can easily be traced back with a traceability system. Moreover, in case of quality-related issue and product recall initiation from market, the actual responsible stakeholder can be identified and penalised for the negligence.

(c) **Marketing**: Traceability can also act as marketing tool which can boost consumer confidence in the brand and can help in making informed buying decisions.

(d) **Logistics Management**: With the real-time knowledge of location of individual product through tracking tags, traceability can be helpful for logistics management. Products can be delivered at right time and quantity with proper control on the inventory. Moreover, with the knowledge of raw material composition and other product-related data, traceability can also overcome the challenges of circular supply chain and be helpful for managing reverse logistics operation.

(e) **Supply Chain Circularity**: Traceability can be a useful tool to overcome challenges of circular supply chain. It can automate collection, sorting and segregation process and help in recycling process of the product with the knowledge of the actual raw material composition.

Let us take some case scenarios of traceability. How do you know the cotton is not made within the legitimate factory? How can you make sure each part used in textile is authentic? How do you know which article is real or fake? How can you track down the complete history of an article? There are many questions including the above-mentioned, which rely on how transparent and accessible is the traceability data. In order to be certain and to make this complete cycle transparent and traceable, it is important to have a system which authenticates the traceability information such as logs of all transactions and events of the life for each individual item in the supply chain. These online system-based ledgers can store the complete end-to-end history of the authentic material, the complete life cycle of the product until the delivery of the product. In this context, one option is that each supply chain actor implements its own supply chain management system with an isolated secure database where the owner has control on the information access provided to other supply chain actors. Practically, this system can make trust up to certain levels, but when multiple parties will involve in a transaction, there might be an issue with trust and transparency. This is due to the reason that the data is stored at one vendor location or vendor-managed database; therefore, he has full control on data and, in fact, the data can be tempered or fabricated. Therefore, in this scenario, there is a lack of trust in terms of data authenticity. To overcome this issue, each vendor can create its own ledger, which can be accessed by other vendors or supply chain stakeholders. In this case, there might have discrepancies in multiple databases as each vendor may have their own life cycle of data and standards, which can mismatch with another vendor database. Again, one stakeholder may not trust the data present on another stakeholder database as one can temper the data and no central authority is there to validate its authenticity. Furthermore, the complexity increases when you are tracing articles for a supply chain where multiple actors may have different standards of maintaining records which can or cannot be trusted. Therefore, traceability is not just recording the information, but it is also equally important to authenticate the information and ensure that each maintains the required standards for data record and exchange. Since there are multiple actors, it is important to link the data of multiple actors to create a complete traceability network.

4 What Is Blockchain and How It Differs from Regular Digital Ledger?

Widely known as the technology behind the Bitcoins and other cryptocurrencies, blockchain is a digital ledger comprising of blocks of information (Wang et al. 2017). Once pre-estimated numbers of transactions are recorded, these blocks of information are added to a ledger; thus, these blocks are linked together to create a chain of blocks called blockchain (Tian 2016).

A blockchain can be conceptualised as an evolutionary digital jigsaw, which consists of the digital blocks. Each block is unique and formed by taking neighbouring blocks into account. It consists of hash values—also known as hash codes—of neighbouring blocks, combined with its own data and then mapped into a new hash value. The mapping is carried out using a function—known as hash function—which is an important building block of a blockchain. It is a mathematical function, which scrambles or encrypts the input data into an output (Abeyratne and Monfared 2016). There appears no apparent relation between the inputs and the outputs without knowing the exact hash function. Therefore, it is computationally extremely difficult to predict the hash function with hit-and-trial or other methods.

Since each block of a blockchain uses the hash value of other blocks, all blocks are interdependent to each other. In fact, this contributes towards the security against tempering of blockchain system. The hash technology and openness or decentralised control and validation of information are the two key characteristics that distinguish the blockchain technology and make it more secured than the conventional digital ledgers. With the hash technology, each new block is encoded with a hash that is generated arithmetically utilising the data content of the block and the hash of the previous block. Therefore, modification or tampering of information within any block will not only change the hash value of that particular block, but the hash of each block afterwards would change and require re-computing. Furthermore, the blocks need to be validated by multiple parties involved in the blockchain implementation; therefore, changing the data after initial creation of the hash is not easy and can be easily detected during the validation process by other parties (Abeyratne and Monfared 2016). Due to the open system and involvement of individuals, generating and validating hash codes in distributed ledgers create a high degree of trust. These novel characteristics encourage the use of blockchain in revolutionising the businesses. As each transaction is recorded with time-based function and visible, it can be tracked and retrieved at any point of time.

5   Traceability in the T&C Supply Chain and Blockchain

Traceability is an ability to identify and trace the data related to a product. The length and breadth of the data depend on how deep or precise traceability information is required. For instance, traceability can be as simple as knowing the origin of raw materials used to produce a shirt; at the same time, it can be precise, for example, the quality data of each raw material used in the manufacturing, the time stamps of material flow within the supply chain, various actors involved in manufacturing and distribution, etc. Moreover, multiple actors are involved in the supply chain with each having certain responsibility of processing or handling the raw or intermediate materials to form the final product. Therefore, a complete traceability can be established only by linking each actor so that the data related to a traceable resource unit or TRU (e.g. final product, intermediate material batch). In this context, one of the fundamental aspects of the traceability is to ensure that each actor maintains the agreed data related to the TRU and links it with other supply chain actors (Kumar et al. 2017b). In order to avoid the external influence, it should be an internal practice that each supply chain actor captures and stores the required data related to a TRU. However, in this vicinity, there should be an automatic authentication process, which ensures that each actor records the required data.

The concept of blockchain is very similar to creating links for data. It should be noted that blockchain is not a database; rather, it is an open ledger where the entries are linked in a mathematical way. The linking not only allows creating the path among a pool of entries but also ensures that if there is any manipulation, it is reflected automatically. As each transaction in the blockchain is recorded in the form of hash code showing which parties were involved, transaction detail and timestamp with a digital signature of authentic party or vendor wherein supply chain each article or good is represented in the form of unique serial number, barcode or tag represented in physical form. Once blockchain encrypts this data and sends it to all peers for verification. Once the other peers "accept" the changes, the block is added to the ledger. This connection provides an audit trail for each transaction in the network so that if there is any unauthorised change in any block, other peers will either validate the change as "valid" or "reject" the change based upon written smart contracts (business logic) (Abeyratne and Monfared 2016). Now, the data is stored on public ledger over the permissioned network so that one can track the chronological history of good and no one can cheat because everyone is watching each transaction over the network through consensus. This way blockchain will store complete immutable network history, and now, two or more parties can share data with each other without involving the third party to create trust (Francisco and Swanson 2018). However, blockchain was initially related to a cryptocurrency known as Bitcoin for financial transactions (Kennedy et al. 2017), where the idea was to create a public ledger for financial transaction, which is visible to all, and everyone can participate in it. On the other hand, for managing traceability, unlike cryptocurrency—where anyone can participate in blockchain network—every actor in the supply chain has specific duties and the public should not alter it. Therefore, traceability needs a permissioned blockchain implementation to create the distributed ledger, where only predefined actors can participate, contribute and access the information (Abeyratne and Monfared 2016; Francisco and Swanson 2018). Therefore, it requires a blockchain network with customisable privacy level. In a permissioned network, only authenticated user can add or edit data into the blockchain and other authenticated user can view the data only. At an enterprise level, sometimes some sensitive information need to isolate from different parties. For example, one vendor X is selling the cotton to buyer A and buyer B at different prices. Since they all are using a common distributed ledger, vendor X would like to hide the buyer's A pricing from buyer B and vice versa. To keep such sensitive information safe and protected, different channel can be created using smart contracts. Nonetheless, in a T&C supply chain, all supply chain actors act as nodes on the network, who handle the materials; therefore, it directly or indirectly contributes to one or more parts of traceability information. Therefore, each participant can have the same copy of distributed ledger. When one party does one transaction (or changes state of record or contributes to the traceability information), this will be updated in all replicated copies of ledger (distributed ledger). When everyone is the using same distributed ledger, it reduces the time to trace the end-to-end history of an article. Querying distributed ledger can trace product virtually where authentic input was provided by permissioned users, but supply chain traceability also has scope to introduce IoT to trace product physically using IoT where things can be tracked down and realtime data can be inserted into the same ledger so that vendor can make sure the virtual and physical presence of materials or products are same.

6 Use Case Example

The following use case conceptualises the implementation of the blockchain in the textile supply chain. Here, it shows a supply chain consisting of eight actors, starting from cotton fibre managing industry to the disposal of the product. Further, it includes the third parties required for the managing the blockchain and their audits. The following are the brief summary of the involved actors or activity:

- Cotton producer—produces cotton or collects cotton fibres.
- Manufacturer—the industries involved in the processing of the cotton fibre to produce final products like shirt or towel. In a typical textile supply chain, manufacturer includes yarn spinning industries (convert fibres into the yarns), weaving mills (produces fabrics from the individual yarns), industries dealing with chemical finishing and dyeing and garment manufacturing houses.
- Distributor—distributes the final product to the retailers.
- Retailer—seller of the final product to the end customers.
- Customer—end-users of the product.
- Disposal—supply chain stakeholder dealing with the disposal of the product, such as product disposal or recycling.
- Registrar—the admin of the blockchain network that can create the credentials of supply chain actors and set their privileges.
- Auditors—to audit quality and authenticity of product, certify the product.

Since there are multiple actors involved, the architecture for the implementation of the blockchain needs to have multiple modules as defined below (Fig. 2) and each module can be deployed on blockchain independently. Nonetheless, they need to interact with each other as they are on the same blockchain.

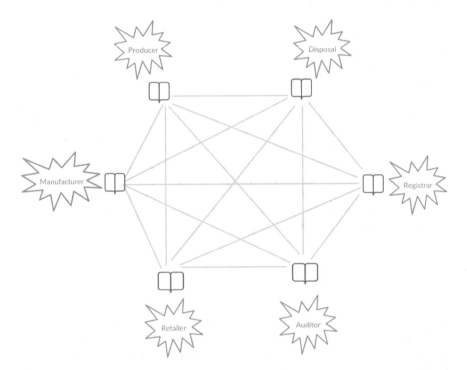

Fig. 2 Blockchain modules

Registration Module: Each node in blockchain acts as an actor, and each actor needs to register through trusted registrar with the blockchain network before accessing it by providing their digital identity as a public certificate to the registrar. This ensures that only authorised supply chain actors can access the supply chain.

Producer Module: Producer is probably the first actor from where the traceability starts. Therefore, it is important that the producer ensures and provides enough authentication that the properties of the produced material are correct. In this context, thus, module certifies the production capacity, production dates, taxonomy and sales registration of good with the help of unique tag to enter data into the blockchain.

Manufacturing Module: Manufacture delivers the final product using the data used for raw material in blockchain and makes sure that input material must be used to produce final output and again blockchain will provide an audit of final physical product produced using raw material should match the input capacity. Now, manufacturer needs to create the unique identity of the products like barcodes, serial numbers and digital tags (e.g. RFID, NFC, genetic tags) and update this data into the blockchain.

Retailer/Customer: After manufacturer tagging, it is also very important that the retailer or customer is getting original product. Therefore, this module provides an interface to trace the complete history of product using tag attached with the product. This module acts as a reading module for the authentication since the retailer neither applies any operation on the product nor changes its characteristics to add traceability information.

Auditor/Certifier: The auditor module is an important component of the blockchain. Using this module, auditor/certifier can authenticate that the final product and input materials (blockchain data and physical material) are correct and there exists a mass balance (i.e. the amount of production is equivalent to the input materials). If there is any discrepancy, the auditor/certifier can raise the flag. Blockchain has advantage of fast settlement as all transactions are on same network. Therefore, this module maintains the authenticity of the information being circulated in the supply chain using blockchain.

Disposal: Anyone can scan the attached tag before handing over the product to recycling so that its status can be flagged as "disposed" in the blockchain.

7 Limitations of Blockchain-Based Traceability System

Cost: Cost can be a major drawback for blockchain technology in the current scenario. Due to its complex network with involvement of numerous transactions, cost for maintaining the blockchain-based system can be high (Tian 2016). Decentralised signature verification can be computationally complex and can be a bottleneck, especially for the T&C supply chain with thousands of products and related information per season. Moreover, millions and trillions of transaction verifications use subsequent amount of computer power and energy.

Immaturity of System: Blockchain technology is still developing and at nascent stage (Tian 2016). Developers and scientists still experimenting the beta level of the technology to solve the challenges related to speed, verification and data limitations.

Security at Product Level: To make a traceability system secure from attack, security mechanism must be installed at two levels (Azuara et al. 2012): One is business or information level, where the traceability information is recorded and shared among the businesses or the supply chain actors. They need to ensure flow of secure and reliable information mainly at business-to-business level. Second is product level, where the product and the attached traceability marker (in form of traceability tags such as RFIDS, barcodes) needs to be secured from attack in form of counterfeits. In such case of tag or product replication, information related to the authentic product can be accessed from the product database even by the counterfeit product. Therefore, in order to have a complete secured traceability system, security must be ensured at product level with the use of a secured traceability marker (Agrawal et al. 2017).

Block Complexity (Size and Number): With the increase in information, supply chain partners and product complexities, the number and size of block will increase, resulting in larger database and computational requirements (Yli-Huumo et al. 2016). To handle such situations, only important product life cycle stage and related relevant information must be explored and selected for traceability purpose.

Human Error: Possibilities of human error cannot be neglected, especially in case of traceability database development and maintenance (Rouse and Rouse 1983; Wang et al. 2017). For traceability, the required data must be recorded and stored at each level of supply chain. Any bad quality information can cause a chain reaction and affect the whole blockchain.

Information Collection: Collecting and recording good quality information is a bigger challenge in the complex T&C supply chain (Kumar et al. 2017a). Actors at various supply chain stages are reluctant to share all the crucial information due to fear of competition, as they believe more transparency and openness will reveal their trade secret to their competitors. On the other hand, information collection and recoding can be a tedious process and lack of motivation unless compensated by any incentive.

Technology Integration: Blockchain adaptation and integration into the existing system can be a major challenge (Wang et al. 2017). It might require significant changes or complete technology replacement that can be costly and time-taking process. Therefore, high motivation and willingness among the involved actors are required to implement the technology. Moreover, blockchain adaptation will also be a cultural change from the centralised to decentralised network that might require investment in human resources.

8 Conclusions

Traceability is a growing concern in the T&C supply chain for creating visibility and transparency. There are multiple challenges in the implementation of traceability, among which one challenge is maintaining the trail of information at different level of supply chain. In this context, as blockchain provides an opportunity to create a link of information network, it has a great potential in implementation of traceability. This chapter provides an overview on how blockchain can be used in implementing traceability, which can overcome the limitations of the conventional way of connecting information. However, blockchain requires the infrastructure and skilled work force to handle it, whereas the T&C supply chain consists of multiple actors, who may not have the means to implement or skilled manpower to maintain it. Nonetheless, the implementation of the blockchain can create a better transparency, traceability and accountability of the supply chain actors.

References

Abeyratne SA, Monfared RP (2016) Blockchain ready manufacturing supply chain using distributed ledger. https://doi.org/10.15623/ijret.2016.0509001

Agrawal TK, Koehl L, Campagne C (2017) Implementing traceability using particle randomness-based textile printed tags. IOP Conf Ser Mater Sci Eng 254:072001. https://doi.org/10.1088/1757-899X/254/7/072001

Agrawal TK, Koehl L, Campagne C (2016) Cryptographic tracking tags for traceability in textiles and clothing supply chain. In: Uncertainty modelling in knowledge engineering and decision making. World Scientific, pp 800–805

Azuara G, Tornos JL, Salazar JL (2012) Improving RFID traceability systems with verifiable quality. Ind Manag Data Syst 112:340–359. https://doi.org/10.1108/02635571211210022

Bhargava B, Ranchal R, Othmane LB (2013) Secure information sharing in digital supply chains. In: 2013 3rd IEEE international advance computing conference (IACC), pp 1636–1640

Caniato F, Caridi M, Crippa L, Moretto A (2012) Environmental sustainability in fashion supply chains: an exploratory case based research. Int J Prod Econ 135:659–670. https://doi.org/10.1016/j.ijpe.2011.06.001

Caridi M, Crippa L, Perego A, Sianesi A, Tumino A (2010) Do virtuality and complexity affect supply chain visibility? Int J Prod Econ 127:372–383. https://doi.org/10.1016/j.ijpe.2009.08.016

Corbellini S, Ferraris F, Parvis M (2006) A cryptographic system for brand authentication and material traceability in the textile industry. In: 2006 IEEE instrumentation and measurement technology conference proceedings, pp 1331–1335

De Brito MP, Carbone V, Blanquart CM (2008) Towards a sustainable fashion retail supply chain in Europe: organisation and performance. Int J Prod Econ 114:534–553. https://doi.org/10.1016/j.ijpe.2007.06.012

Doorey DJ (2011) The transparent supply chain: from resistance to implementation at Nike and Levi-Strauss. J Bus Ethics 103:587–603. https://doi.org/10.1007/s10551-011-0882-1

Ekwall D (2009) The risk for detection affects the logistics system setup for cargo smugglers

Francisco K, Swanson D (2018) The supply chain has no clothes: technology adoption of blockchain for supply chain transparency. Logistics 2:2

Grimm JH, Hofstetter JS, Sarkis J (2016) Exploring sub-suppliers' compliance with corporate sustainability standards. J Clean Prod 112:1971–1984. https://doi.org/10.1016/j.jclepro.2014.11. 036

Kennedy ZC, Stephenson DE, Christ JF, Pope TR, Arey BW, Barrett CA, Warner MG (2017) Enhanced anti-counterfeiting measures for additive manufacturing: coupling lanthanide nano-material chemical signatures with blockchain technology. J Mater Chem C 5:9570–9578. https://doi.org/10.1039/C7TC03348F

Kumar V (2017) Exploring fully integrated textile tags and information systems for implementing traceability in textile supply chains

Kumar V, Agrawal TK, Wang L, Chen Y (2017a) Contribution of traceability towards attaining sustainability in the textile sector. Text Cloth Sustain 3:5. https://doi.org/10.1186/s40689-017-0027-8

Kumar V, Hallqvist C, Ekwall D (2017b) Developing a framework for traceability implementation in the textile supply chain. Systems 5:33. https://doi.org/10.3390/systems5020033

Kumar V, Koehl L, Zeng X, Ekwall D (2017c) Coded yarn based tag for tracking textile supply chain. J Manuf Syst 42:124–139. https://doi.org/10.1016/j.jmsy.2016.11.008

OECD (2017) OECD due diligence guidance for responsible business conduct responsible supply chains in the garment and footwear sector

Peres B, Barlet N, Loiseau G, Montet D (2007) Review of the current methods of analytical trace-ability allowing determination of the origin of foodstuffs. Food Control 18:228–235. https://doi.org/10.1016/j.foodcont.2005.09.018

Report M (2017) Marketline report store—global textile mills. Marketline, London, UK

Rouse WB, Rouse SH (1983) Analysis and classification of human error. IEEE Trans Syst Man Cybern SMC 13:539–549. https://doi.org/10.1109/tsmc.1983.6313142

Tachizawa EM, Wong CY (2014) Towards a theory of multi-tier sustainable supply chains: a sys-tematic literature review. Supply Chain Manag Int J 19:643–663. https://doi.org/10.1108/SCM-02-2014-0070

Tian F (2016) An agri-food supply chain traceability system for China based on RFID blockchain technology. In: 2016 13th international conference on service systems and service management (ICSSSM), pp 1–6

Wajsman N, Arias Burgos C, Davies C (2015) The economic cost of IPR infringement in the clothing, footwear and accessories sector. European Union Intellectual Property Office, Alicante, Spain

Wang J, Wu P, Wang X, Shou W (2017) The outlook of blockchain technology for construction engineering management. Front Eng Manag 4:67–75. https://doi.org/10.15302/J-FEM-2017006

Yang Y, Martin W, Yanagishima K (1997) Evaluating the benefits of abolishing the MFA in the Uruguay Round package. In: Global trade analysis: modeling and applications. Cambridge University Press, pp 253–279

Yli-Huumo J, Ko D, Choi S, Park S, Smolander K (2016) Where is current research on blockchain technology?—A systematic review. PLoS ONE 11:e0163477. https://doi.org/10.1371/journal.pone.0163477

Part III
AI for Garment Design and Comfort

Artificial Intelligence Applied to Multisensory Studies of Textile Products

Zhebin Xue, Xianyi Zeng and Ludovic Koehl

Abstract Multisensory evaluation is an interesting realm in sensory study. It deals with relations between not only different human sensory modalities but also different levels of perceptions. As for textile products, multisensory research investigates textile or apparel properties perceived through different senses as well as the relations between sensory factors and perceptions of higher levels such as preference and emotion. In order to investigate complex relations between different sensory datasets, many analytic and computing methods are available. Compared with statistics, or classical computing methods, as perhaps the most often used data mining methods, artificial intelligent tools, due to their higher capacity in handling data uncertainty and imprecision, have been showing more advantage in dealing with human-related knowledge which is representative in sensory studies. In this chapter, we are going to discuss the recent application of artificial intelligence to the study of fabric tactile properties from a multisensory point of view. To be specific, the whole work is divided into two major parts. In the first part, the intelligent tools of fuzzy comprehensive evaluation and genetic algorithm have worked together to study the relations between fabric tactile properties and the product's total preference as perceived by consumers. The second part is about the visual interpretation of fabric tactile properties in a virtual environment through a systematic method based on rough inclusion degree and fuzzy inference systems.

1 Novel Sensory Methodologies for Fabric Hand Study

Clothes are a kind of sensory item. Consumers' purchasing decision is made upon their sensory experience about the product. In recent years, much effort has been

Z. Xue (✉)
School of Textiles and Clothing, Jiangnan University, Wuxi, China
e-mail: zhebin.xue@hotmail.com

Z. Xue · X. Zeng (✉) · L. Koehl
GEMTEX, Ecole Nationale Supérieur des Arts et Industries Textiles, Roubaix, France
e-mail: xianyi.zeng@ensait.fr

© Springer Nature Singapore Pte Ltd. 2018
S. Thomassey and X. Zeng (eds.), *Artificial Intelligence for Fashion Industry in the Big Data Era*, Springer Series in Fashion Business,
https://doi.org/10.1007/978-981-13-0080-6_11

dedicated to the study of sensory properties of textile products from both theoretical and industrial point of view. According to the definition of sensory evaluation initially put forward by the Institute of Food Technologists, the sensory study of textile products can be considered as a scientific discipline used to evoke, measure, and interpret reactions to those characteristics of textile materials as they are perceived by the senses of sight, smell, taste, touch, and hearing (Dijksterhuis 2008). Research works can be found in almost all the above-mentioned sensory aspects, but among which the feeling of touch or fabric hand has been attached highest importance as a crucial appraisal of quality and prospective performance of an apparel product. Since fabric hand is closely related to the textile mechanical properties and meanwhile an important index affecting consumers' experience about the overall comfort of a product, it is the bridge communicating supply and demand of textile market and has been extensively researched from every possible angle in relevant domains (Philippe et al. 2004; Jeguirim et al. 2010; McGregor et al. 2015).

Normally, a sensory evaluation is carried out by one or several panels of trained assessors according to standardized descriptors, techniques, and procedures (Stone et al. 2012; ISO 8586 1993; ISO 11035 1995). It requires careful design of experiments and strict control of implementation to ensure the reliability of obtained sensory data.

Based on the data obtained from sensory experiments, the next important step is to find a proper way to process the sensory data to finally produce effective and interpretable results. To date, many data mining methods have been developed to explore complex relations between different datasets. The most commonly used methods are based on statistics, including principal component analysis (PCA) (Jolliffe 2010), linear regression analysis (Weisgerg 2005), multiple factor analysis (Howorth et al. 1958; Le Dien 2003), multidimensional scaling (Hollins et al. 1993; Picard et al. 2003), and various kinds of correlation coefficient analysis (Härdle and Simar 2010). These methods are efficient in solving a lot of characterization and modeling problems due to their high capacity in recognizing linear patterns lying beneath different sources of information and then discovering correlations between data and between attributes from a large quantity of numerical data (Giboreau et al. 2001; Suelar et al. 2007; Fernandes et al. 2008).

But most real-world problems are complex and contain uncertain relations. The assessment of sensory properties of textile products is such a representative example that people often base their feeling about an product such as fabric hand on their senses and knowledge and express it in ambiguous manners such as words, which is hard to quantify and thus impossible to manipulate. Classical quantitative methods are found to be less competent. First, as long as a problem is dealing with human knowledge, the concerned relations are often nonlinear. The application of statistical techniques might cause important information loss due to their linearly structured models. Next, there often exists high uncertainty and imprecision in data structures due to non-unified linguistic human evaluation scores. But most of the classical analysis methods can only process perfect and complete numerical data without any uncertainty or imprecision. In addition, classical methods cannot always lead to precise and significant physical interpretation of data. Finally, classical methods often have strict requests on the size and distribution of the database. But to collect a big number of human evaluation data is quite time-consuming and not practical for many researches, for example in pilot studies. With a limited collection of samples, it is unlikely to produce good fit modeling results via classical methods.

In this situation, in recent years, intelligent computational techniques, such as artificial neural network (ANN) (Fausett 1994), fuzzy logic (Zadeh 1965; Sugeno et al. 1993), GA (Goldberg 1989), and many hybrid applications of these tools (Ruan 1997) have been extensively applied to the modeling and analysis of physical and human data (Park et al. 2000; Zeng and Koehl 2003; Zeng et al. 2008). Compared with classical methods, they are capable of (1) dealing with both numerical and linguistic data, (2) solving nonlinear problems, (3) modeling human expert reasoning so as to produce precise and straightforward interpretation of results, and (4) computing with relatively small sets of data and without need of any preliminary or additional information like probabilistic distributions in statistics.

2 Prediction of Emotional Preference from Fabric Tactile Properties Based on a Fuzzy-Genetic Model

As defined by the Textile Institute in 1975, fabric hand (or fabric tactile properties) is the 'Subjective assessment of the textile material obtained from the sense of touch' (Ali and Begum 1994). Fabric hand is regarded as an identity of a fabric. It determines to a large extent consumers' general preference toward a textile product. Exploring the tactile dimensions of a textile product and further discussing their relations with consumers' emotional preferences, is of great significance to textile researchers and manufacturers to develop "right" products that are better accepted by the market.

In the current research, sensory experiments are carried out by a panel of experts on a set of suiting fabrics according to standardized sensory techniques and procedures. In our problem, both tactile properties and emotional preferences are sensory information which involve much uncertainty and imprecision. Thus in our study, a new approach based on the intelligent tools such as fuzzy sets theory and GA was proposed to investigate the relations between different sensory datasets. The method of fuzzy comprehensive evaluation (FCE) is employed to model the relations between the tactile properties of the samples and the general preference with respect to the end-use of men's suit. The GA is applied to quantitatively measure the impact of each sensory attribute on the total preference by producing a near-optimal weight distribution among the sensory variables. Finally, the efficacy of the proposed fuzzy-genetic model is verified by comparing the results obtained from the model and those from the sensory experiments.

2.1 Sensory Experiments on Suiting Fabrics

2.1.1 Experimental Materials

Fifty textile fabrics with various tactile properties were selected as the experimental samples. Since here we take men's suits as the research object, the materials that are commonly used for producing men's suits, such as wool, cotton, polyester, and their blends, are considered in our study.

Samples were pre-cut into standardized dimensions (300 mm × 300 mm) and then labeled with random codes to prevent panelists from being influenced. Before the tests, all samples were conditioned for a minimum of 24 h under standard atmospheric conditions (20 ± 2 °C temperature, $65 \pm 2\%$ relative humidity). The sensory experiments were completely performed under these conditions.

2.1.2 Sensory Experiments to Obtain Attributes on Fabric Hand and Total Preference

(1) Determination of sensory descriptors

Sensory descriptors are verbal terms used during sensory evaluations for describing the human perceptions on products. In the current study, a sensory experiment was organized to extract tactile properties and the total preference of a set of suiting fabrics. In this study, a normalized procedure has been taken to generate fabric tactile properties according to the current sample base. After collecting, screening, and adjusting, seven representative fabric tactile attributes have been determined. Table 1 shows some examples (D1 and D2) of the selected descriptors along with the definition and assessing techniques related to each of them. In addition to the seven descriptors, a preference index, 'Total preference' has been defined, with respect to which the predictive mechanism of fabric tactile properties would be modeled.

(2) Sensory evaluation scales

For each descriptor, an eleven-point scale degreed from 0 (no attribute detected) to 10 (upper limit value of the attribute) was used. For example, if a sample was considered by a panelist as neither soft nor stiff, then the value of softness for the specific fabric would be 5. Meanwhile, a seven-point scale degreed from 1 to 7 was used for the evaluation of the total fabric hand.

(3) Panelists

Sensory evaluations were carried out by a panel of 15 female and 8 male subjects between 21 and 45 years. They are experts with a textile background such as professors, Ph.D. students, and postgraduate students who work or study in a college of textile and clothing engineering, and professionals who work mainly on fabric design and suit production.

Table 1 Definitions and assessing techniques of sensory descriptors

Descriptor	Definition	Assessing technique
D1-Softness	Resistance/non-resistance to compression or bending	Fabric is held between the thumb and the other four fingers of one's most used hand. While moving the fabric back and forth, one assesses the resistance
D2-Smoothness	The amount of small particles rising on fabric's surface. The surface of a smooth fabric will offer little resistance to slipping when rubbed	To move the palm of the hand across the surface of the sample
…	…	…
T-Total preference	The general preference for the end-use of men's suits	Free gestures as long as is needed

The panelists were required to take a 2 h per week training which lasts 6–8 weeks to get adequate knowledge about the touch of fabrics, the descriptive terminology, the techniques and procedures of sensory evaluation, etc.

(4) Control fabrics

The purpose of using the so-called control fabrics is to avoid significant disagreement among the panelists, which can to some extent improve the reliability of the obtained sensory data. They were selected from the testing samples. Every panelist was asked to choose two samples to represent, respectively, the highest (i.e., 10) and the lowest score (i.e., 0) for each tactile attribute. A discussion was organized to reach a general consensus among the panelists. In the end, seven pairs of control fabrics (corresponding to the seven tactile descriptors) were determined to provide a reference to the panelists during the evaluations. Notably, it is not necessary that all the panelists hold exactly the same opinion about the control fabrics.

During the actual experiments, the panelists were required to score the samples by taking into account the results of the control fabrics. Since the panelists were well trained, and the selection and scoring of the control fabrics gained general consensus among the panelists, the use of control fabrics as evaluation reference will not affect the panelists' willing and ability to express their real feeling about the samples in the current sensory experiments.

(5) Sensory evaluation procedure

Sensory experiments were carried out in two scenarios, the tactile scenario and the total preference scenario, to obtain the panelists' tactile perceptions about the samples and their general preferences with respect to the end-use of men's suits. The break between these two scenarios is approximately one week for each panelist.

a. Tactile scenario

Before the experiments, all the samples were conditioned for a minimum of 24 h under the standard atmospheric condition (20 ± 2 °C temperature, $65 \pm 2\%$ relative humidity). Each panelist was asked to wash and dry his/her hands with the non-moisturizing soap and paper towel provided.

The panelists were allowed to both see and touch the fabric during the evaluation, which is in accordance with our real-life experience. The sensory evaluation should be carried out individually for each panelist. Each panelist might start the evaluation whenever he/she was ready. Control fabrics were involved in the real tests. The panelists were free to give scores to the control fabrics, yet based on the general consensus reached previously.

b.   Total preference scenario

Before the evaluation, each sample fabric is prepared in the same way as it was in the tactile scenario. Each panelist was asked to wash and dry his/her hands with the non-moisturizing soap and paper towel provided.

In this scenario, the panelists were asked to give a score to each sample according to the extent to which it is suitable for producing a men's suit. The panelists made their own decision according to their professional or personal experience and knowledge about men's suits.

2.2   Predictive Model Based on a Fuzzy-Genetic Algorithm

In simple words, the predictive model is developed based on the resolution of the inverse problem of fuzzy comprehensive evaluation by a genetic algorithm.

2.2.1   Inverse Problem of Fuzzy Comprehensive Evaluation

(1)   Fuzzy comprehensive evaluation

Comprehensive evaluation is a kind of method used to make a general judgment on a multifactor problem taking into consideration the individual impact of each influential variable. On this basis, fuzzy comprehensive evaluation (FCE) is an extension of the classical comprehensive evaluation method by introducing fuzzy sets principles to better deal with the uncertainty and imprecision often encountered in real-world problems.

FCE as an artificial intelligent method can provide a high level of confidence in decision-making. Similar to judgment by human beings, it would classify or distinguish things by means of analyzing fuzzy information as much as possible (Yager 1977; Zimmermann 2011).

Classically, FCE is carried out according to the following steps.

a. Establishment of the factor set U

Suppose, $U = \{u_1, u_2, \ldots, u_n\}$ is a set composed of n kinds of factors, called the factor set. In this study, the factor refers to the seven ($n=7$) sensory evaluation attributes. Thus, the factor set can be represented as:

$$U = \{\text{Softness, Smoothness, Fullness,}$$
$$\text{Delicacy, Flexibility, Lightness, Resiliency}\}$$

b. Establishment of the weighting set $\underset{\sim}{W}$

It is necessary to establish a so-called weighting set $\underset{\sim}{W}$ where any element w_i ($i = 1, 2, \ldots, n$) quantifies the impact of the corresponding attribute u_i ($i = 1, 2, \ldots, n$) on the total preference evaluation.

The weighting set $\underset{\sim}{W}$ is a fuzzy subset of the factor set U.

$$\underset{\sim}{W} = (w_1, w_2, \ldots w_n)$$

in which the weight w_i represents the membership of the element u_i with respect to the fuzzy set $\underset{\sim}{W}$. We have $\sum_{i=1}^{n} w_i = 1, 0 \leq w_i \leq 1$ ($i = 1, 2, \ldots, n$), and $n=7$ in the current study).

c. Establishment of the evaluation set V

Suppose, $V = \{v_1, v_2, \ldots, v_n\}$ is called the evaluation set which is composed of n possible evaluation results. In the present study, the evaluation set refers to the eleven-point rating scale which can be represented by, $V = \{0, 1, \ldots, 10\}$ or verbally, $V = \{\text{Extremely not}, \ldots, \text{Medium}, \ldots, \text{Extremely}\}$

d. Construction of the fuzzy evaluation matrix $\underset{\sim}{R}$

For an evaluation problem like the one in this study, there exists a fuzzy mapping $\underset{\sim}{f}$ from U to $F(V)$:

$$\underset{\sim}{f} : U \rightarrow F(V), \quad \forall u_i \in U$$

This fuzzy relation can be represented by a fuzzy evaluation matrix $\underset{\sim}{R}$ which is defined as follows:

$$\underset{\sim}{R} = \begin{bmatrix} r_{11} & r_{12} & \cdots & r_{1m} \\ r_{21} & r_{22} & \cdots & r_{2m} \\ \cdots & \cdots & \cdots & \cdots \\ r_{n1} & r_{n2} & \cdots & r_{nm} \end{bmatrix}$$

Here, the expression (U, V, R) is called the FCE model.

e. Fuzzy comprehensive evaluation

If the weighting set W, representing the influence of each sensory attribute on the total preference evaluation, is properly decided, the FCE will be carried out according to the FCE model as follows:

$$B = W \circ R \tag{1}$$

or,

$$(b_1, b_2, \ldots, b_m) = (w_1, w_2, \ldots, w_n) \circ \begin{bmatrix} r_{11} & r_{12} & \cdots & r_{1m} \\ r_{21} & r_{22} & \cdots & r_{2m} \\ \cdots & \cdots & \cdots & \cdots \\ r_{n1} & r_{n2} & \cdots & r_{nm} \end{bmatrix} \tag{2}$$

where fuzzy set B is called the decision set. n is the number of the tactile attributes (factors) $(n=7)$, and m is the number of the possible scores $(m=11)$ in the evaluation set.

In this expression, the symbol 'o' represents the composition operator. In this study, the multiply-add operator (i.e., $(\bullet, +)$) is employed to retain as much information as possible from the sensory relations.

(2) Inverse of FCE

In real problems, we often encounter the situations where the relation matrix R and the decision set B are given while the weight set W is unknown, which is called the inverse problem of FCE. The problem to be concerned in our study is right of this kind, as our aim is to investigate, based on the sensory data obtained from the tactile and total preference evaluation scenarios, to what extent each tactile property has an impact on the total preference of a suiting fabric.

a. Alternative weighting set

The often used method to solve an inverse problem of FCE is based on experts' professional knowledge and experience. In this method, a set of experts of related background are invited to decide by discussion one or several weighting sets which are called the alternative weighting sets $A_i = (a_1, a_2, \ldots, a_n)$ $(n = 7)$. The solution set J is the assembly of the alternative weighting sets. So we have

$$J = \left(A_1, A_2, \ldots, A_s \right)$$

where s refers to the number of solutions, and n is the number of sensory attributes $(n=7)$.

Once the solution set J is determined, the assembly of the corresponding decision sets $D = \left(\underset{\sim}{D}_1, \underset{\sim}{D}_2, \ldots, \underset{\sim}{D}_s \right)$ can be obtained according to the following model.

$$\underset{\sim}{D}_i = \underset{\sim}{A}_i \circ \underset{\sim}{R} \quad (i = 1, 2, \ldots, s) \tag{3}$$

b. Close degree

The optimal weight distribution is the one who makes lowest the close degree of the alternative set D_i to the decision set $\underset{\sim}{B}$.

Thus, it is necessary to define a method to measure this close degree $\left(\underset{\sim}{D}_j, \underset{\sim}{B} \right)$. One commonly used method is the Hamming distance which is defined as follows:

$$\left(\underset{\sim}{D}_i, \underset{\sim}{B} \right) = 1 - \frac{1}{n} \sum_{j=1}^{n} \left| \underset{\sim}{D}_i (u_j) - \underset{\sim}{B} (u_j) \right| \quad (i = 1, 2, \ldots, s) \tag{4}$$

where n is the number of evaluation attributes ($n=7$).

The optimal weight distribution $\underset{\sim}{W}_{opt}$ is decided in this manner.

We have

$$\underset{\sim}{W}_{opt} = \underset{\sim}{A}_i \quad \text{when} \quad \left(\underset{\sim}{D}_i, \underset{\sim}{B} \right) = \max_{1 \leq j \leq s} \left(\underset{\sim}{D}_j, \underset{\sim}{B} \right) \tag{5}$$

where s denotes the number of solutions.

2.2.2 Genetic Algorithm Applied to Solve the Inverse Problem of FCE

(1) Mathematical method

Expertise is helpful to solve many real-world problems, but since it depends a lot on experts' personal experience and preference, it is less capable of producing robust and repeatable results of high precision. In this situation, some mathematical optimization techniques are gaining attention in recent years. In the current study, the intelligent tool of GA is introduced to solve the inverse problem of FCE so as to obtain a near-optimal weight distribution among the tactile attributes with respect to the total preference of the sample fabrics.

GA is an exploratory search procedure which is developed upon the principles of natural evolution and population genetics. It uses multiple concurrent search points called 'chromosomes' which are processed through three genetic operations, i.e., selection, crossover, and mutation, to generate new search points called 'offspring' for the next iteration (Davis 1991).

The GA is particularly capable of deriving approximate solutions for multivariable optimization problems where classical mathematical methods are not so competent.

Fig. 1 An individual
represented by a binary
string

w_1	w_2	w_3	w_4	w_5	w_6	w_7

In the present study, the standard genetic algorithm (SGA) is applied to find a near-optimal weight distribution among different sensory attributes, which is in fact a process of solving an optimization problem.

According to the present study, the resolution of the inverse problem of FCE by SGA goes through the following steps.

STEP 1: Encoding. It is important to represent the weights in a way suitable for applying the genetic operators. In our research, each weight w_i ($i = 1, 2, \ldots, 7$; $w_i \in [0, 1]$) is a continuous variable. The initial ranges of the variables constitute a search space or spatial mesh. Each mesh point is called an individual represented by a binary string which is shown in Fig. 1.

For weight w_i, if it is encoded in k binary bits, then

$$\varepsilon = \left(w_{i(\text{max})} - w_{i(\text{min})} \right) / \left(2^k - 1 \right), \ \forall i \in \{1, 2, \ldots, 7\} \tag{6}$$

where ε_{w_i} is the precision of the results of the algorithm (i.e., w_i), $\left(w_{i(\text{max})}, w_{i(\text{min})} \right)$ is the initial range of variation (or the initial search range) with $w_{i(\text{max})}$ and $w_{i(\text{min})}$ the upper bound and lower bound of w_i, respectively.

If an individual is of n variables and encoded by k binary bits, the search space of the problem is of $\left(2^k \right)^n$ mesh points. In our study, the individuals are encoded by 6 binary bits, given that we have 7 tactile variables ($n=7$), the search space of our problem is of $\left(2^6 \right)^7 = 2^{42}$ dimensions. Thus, the precision of the results will be calculated as follows:

$\varepsilon = (1 - 0)/ \left(2^6 - 1 \right) = 0.016$ (in our study, $[0, 1]$ is the range of variation of the weights, i.e., $w_{i(\text{max})} = 1$, and $w_{i(\text{min})} = 0$).

STEP 2: Initialization. An initial population including a number of individuals (or parents) is selected randomly from the above mesh points.

STEP 3: Fitness evaluation.

The major objective of this study can be simplified as to find an optimal $\underset{\sim}{W}_{opt}$ that can maximize the (D_i, B). Here, the Hamming approach is used to measure the close degree or the fitness. Since the GA can only solve minimization problems, the Hamming approach is modified as the following optimization criterion.

$$f_i = \frac{1}{l} \sum_{j=1}^{l} \left| \underset{\sim}{D}_i \left(u_j \right) - \underset{\sim}{B} \left(u_j \right) \right| - 1 \quad (i = 1, 2, \ldots, t) \tag{7}$$

where f_i is the optimization criterion, t refers to the ith parent individual, and l is the number of fabric samples ($l=50$). The lower the unfitness value, the more adaptable is the individual.

On this basis, the fitness function is defined as follows:

$$F_i = \frac{1}{f_i^2 + 0.001}, \quad (i = 1, 2, \ldots, t) \tag{8}$$

where F_i is the fitness value of the ith parent individual.

STEP 4: Proportional selection. In this process, a new generation of population is formed by choosing individuals from the current population according to the proportional function as follows:

$$p_i' = \frac{F_i}{\sum_{i=1}^{t} F_i} = \frac{\frac{1}{f_i^2 + 0.001}}{\sum_{i=1}^{n} \frac{1}{f_i^2 + 0.001}} \quad (i = 1, 2, \ldots, t) \tag{9}$$

where p_i' refers to the selective proportion for the ith individual, t is the number of parent individuals.

Let $p_i = \sum_{k=1}^{i} p_k'$, and $\sum_{i=1}^{n} p_i = 1$, $(i = 1, 2, \ldots, t)$. Obviously, [0, 1] is divided by $\{p_i \,|i = 1, 2, \ldots, n\}$ into n subintervals corresponding, respectively, to n parents. A set of random numbers $\{u(k)| k = 1, 2, \ldots, n\}$ are then generated. If $u(k) \in (p_i - 1, \ p_i]$, then the individual i is selected. In the same manner, another set of random numbers are produced, then two sets, each of n individuals, are selected for the following operations.

STEP 5: Crossover. Under this operation, individuals selected from Step 4 are randomly matched in pairs. In each pair, the two parent strings exchange portions of their structures according to the crossover proportion p_c to produce two new offspring. The higher the crossover proportion, the faster the updating frequency of the population is. In this study, we take $p_c = 0.7$ as a moderate choice. It is believed that two-point crossover is more effective than single-point crossover. Thus, the two-point crossover is adopted here.

STEP 6: Mutation. Mutation is to alter the values of two random positions in some gene strings ('1' to '0,' or '0' to '1') according to the mutation rate p_m. Generally, small values of p_m are more favorable to prevent the disruption of the solutions. Thus, in the current study, $p_m = 0.001$ is adopted.

STEP 7: Evolutionary iteration. The n individuals derived from Step 6 are regarded as the new parents and have to return to Step 3 to participate in a new round of operations of evaluation, selection, crossover, and mutation. In this study, it is determined that the iteration ceases when no significant improvement of the fitness value is detected.

(2) Modeling results and discussion

During the data analysis, we adopt the mean values of each fabric sample on the 8 sensory variables (7 tactile descriptors (D#) and the total preference (T)), which is partially listed in Table 2.

After normalization, the decision set B and the fuzzy evaluation matrix R are obtained for the construction of the FCE model.

Table 2 Mean value of each sample on each sensory descriptor

Sample	Descriptor							
	D1	D2	D3	D4	D5	D6	D7	T
#1	1.95	6.23	5.05	2.59	5.64	3.05	2.32	2.64
#2	7.36	5.73	5.55	5.77	6.77	5.59	5.27	4.50
...	...	...	...	...	...	...	...	...
#50	7.18	7.23	3.73	7.59	7.55	6.91	6.23	5.73

$$\underset{\sim}{B} = \begin{bmatrix} 0.21, \ 0.61, \ 0.33, \ 0.16, \ 0.20, \ \ldots, \ 0.54, \ 0.47, \ 0.50, \ 0.50, \ 0.88 \end{bmatrix}$$

$$\underset{\sim}{R} = \begin{bmatrix} 0.05 & 0.85 & 0.65 & 0.72 & 0.69 & \cdots & 0.09 & 0.25 & 0.55 & 0.31 & 0.82 \\ 0.71 & 0.64 & 0.22 & 0.65 & 0.45 & \cdots & 0.36 & 0.49 & 0.59 & 0.62 & 0.86 \\ 0.44 & 0.51 & 1.00 & 0.03 & 0.13 & \cdots & 0.44 & 0.32 & 0.32 & 0.30 & 0.25 \\ 0.14 & 0.60 & 0.34 & 0.55 & 0.43 & \cdots & 0.17 & 0.18 & 0.45 & 0.39 & 0.86 \\ 0.49 & 0.69 & 0.09 & 0.41 & 0.28 & \cdots & 0.48 & 0.46 & 0.52 & 0.55 & 0.83 \\ 0.14 & 0.53 & 0.03 & 1.00 & 0.70 & \cdots & 0.20 & 0.36 & 0.42 & 0.35 & 0.73 \\ 0.19 & 0.65 & 0.47 & 0.37 & 0.39 & \cdots & 0.27 & 0.31 & 0.46 & 0.32 & 0.81 \end{bmatrix}$$

a. Comparison between weightings obtained from experts and GA approach

As a verification test, n panelists ($n=4$) who are experts recruited from textile faculty or industries with adequate background in the design and production of suits as well as fabric hand evaluation techniques are invited to give alternative weighting sets $\underset{\sim}{A}_i$ ($i = 1, 2, 3, 4$), which are shown in Table 3.

The weight distributions advised by experts are based on their professional expertise. But according to the FCE model and the Hamming Distance, the quality of those results (the close degree ($\underset{\sim}{D}_j$, $\underset{\sim}{B}$)) was not that satisfactory. For example, the close degree of the third solution $\underset{\sim}{A}_3$ is 67.15%, which seems the best of all, while the performance of other solutions is still worse. A particular low close degree of less than 50% is detected on the alternative set $\underset{\sim}{A}_2$.

Also from Table 3, when the GA is operated at the 5th generation, the close degree or fitness value reaches 90%. As the computation continues, the results are constantly improved till the 38th generation. From the 39th generation, the performance tends to be stable, and no further significant optimization has been made on the weight distribution, which indicates that the stop point of the algorithm was approaching. After a little improvement takes place at the last weight, the GA stalled at the 51st generation, when the fitness value exceeds 95%. Compared with the weightings from the experts, the GA approach is verified to be much better at seeking a near-optimal weight distribution among various variables.

Figure 2 shows the final results by the GA approach. It is obvious that among the seven sensory descriptors, descriptor 'flexibility' (D4) has the largest effect on the

Table 3 Weighting results and close degree obtained from experts and GA

		D1	D2	D3	D4	D5	D6	D7	Close degree $\left(\underset{\sim}{D}_j , \underset{\sim}{B} \right)$
Expert advice	$\underset{\sim}{A}_1$	0.1	0.3	0.3	0.5	0.2	0.1	0.1	0.6407
	$\underset{\sim}{A}_2$	0.1	0.3	0.2	0.1	0.3	0.3	0.5	0.4583
	$\underset{\sim}{A}_3$	0.2	0.1	0.3	0.2	0.5	0.1	0.2	0.6715
	$\underset{\sim}{A}_4$	0.3	0.2	0.7	0.2	0.3	0.1	0.1	0.6028
GA results	5th	0.2751	0.0179	0.1522	0.0339	0.4130	0.0198	0.4692	0.8915
	39th	0.0538	0.0179	0.3964	0.6719	0.1626	0.0198	0.1022	0.9256
	51st	0.0538	0.0179	0.3964	0.6719	0.1626	0.0198	0.0416	0.9523

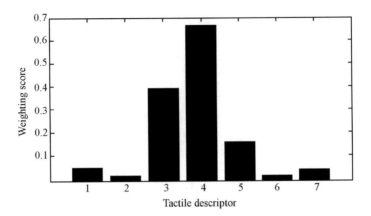

Fig. 2 Final weight distribution by GA approach

total preference of men's suitings, with the score being close to 0.7. The reason could be that fabric flexibility influences the drape of a garment, which is a very important factor for garments' styling. Besides, descriptor 'fullness' (D3) and 'delicacy' (D5) also have significant influence on the total preference evaluation. 'Fullness' is a complex feeling coming from a combination of fabric's thickness, surface roughness, and tactile warmth. Materials with a proper extent of fullness tend to have good shaping effect and are easier to process, without being too bulky or too frivolous. In addition, delicate fabrics (fabrics of higher value on descriptor D5—'delicacy') often leave us a good impression in quality and aesthetic appeal, which is especially the case for formal garments like men's suits. Therefore, 'Flexibility,' 'Fullness,' and 'Delicacy' are found to be the three major factors determining the total preference of a suiting fabric. As regards to the other four tactile attributes, 'softness' (D1), 'smoothness' (D2), 'lightness' (D6), and 'resiliency' (D7), respectively, their impact on the total preference of men's suiting fabrics is relatively small with their weights not exceeding 0.1, thus is regarded as the minor factors.

b. Verification of predictive model

To verify the effectiveness of the proposed fuzzy-genetic model, it is important to compare the predictive results obtained from the model and the experimental results collected from the panelists. According to the FCE model $B = W \circ R$, the weighting set W is a normalized set for which $\sum_{i=1}^{n} w_i = 1$, $0 \leq w_i \leq 1$. But the optimized result obtained from our model $W_{opt} = \begin{bmatrix} 0.0538, 0.0179, 0.3964, 0.6719, 0.1626, 0.0198, 0.0416 \end{bmatrix}$ is not yet normalized. In order to apply the evaluation model, it is necessary to normalize the weighting set W_{opt}, so that we have $\sum_{i=1}^{n} w_{opt(i)} = 1$. The normalization method is defined as follows:

$$w'_{opt(i)} = \frac{w_{opt(i)}}{\sum_{i=1}^{n} w_{opt(i)}} \tag{10}$$

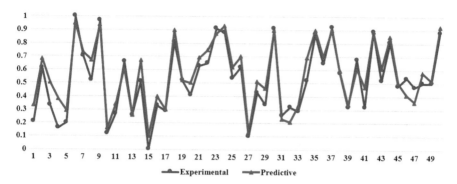

Fig. 3 Comparison between experimental and predictive results

where $w'_{opt(i)}$ is the optimal weight for the ith tactile descriptor after normalization, and $i = 1, 2, \ldots, n$, $n = 7$ in the current study.

After normalization, we have the optimal weighting set W'_{opt} as below.

$$W'_{opt} = \left[0.0394, 0.0131, 0.2906, 0.4926, 0.1192, 0.0145, 0.0305 \right]$$

According to the FCE model and the above optimal weighting set, we obtain the decision set $\underset{\sim}{D}$ or the set of predictive total preference results (represented by the red line) for the 50 samples which are plotted in Fig. 5 as compared with the experimental results (represented by the green line).

As evident in Fig. 3, a good agreement is detected between the experimental and predictive results, with the difference on most fabric samples being relatively small (not exceeding 0.14). For most data pairs, the difference is less than 0.1. It is notable that the experimental results in this research are qualitative scores, which are of high uncertainty and imprecision, to obtain a model that can perfectly predict them is almost unachievable. Therefore, the results obtained in this study are considered as quite satisfactory.

Finally, it can be concluded that the fuzzy-genetic model proposed in this study works efficiently and reliably in interpreting the relations between various tactile attributes and the total preference of men's suitings. For the convenience of observation, the optimal weight distribution $\underset{\sim}{W'}_{opt}$ is simplified by reserving two decimal fractions as follows:

$$\underset{\sim}{W'}_{opt} = [0.04, 0.01, 0.30, 0.50, 0.12, 0.01, 0.03]$$

3 Visuo-haptic Perception of Fabric Tactile Properties Based on a Fuzzy Inclusion Approach

Given the absence of a certain sense, it is natural to consider the possibility of complementing this sense by getting use of the other available senses. This hypothesis is based on our brain's sophisticated mechanism called memory association (Bernard and Juan 2005; Shinomoto 1987; Vogel 2005). It is known that we perceive 80% of the daily information through vision (Wang et al. 2001) and apart from hearing, vision is the only powerful source of information that is available on the Internet. Thus, we intend to ask, in such a non-haptic environment, is it possible to perceive the tactile properties of a textile product by interpreting its visual representations?

The present study aims to investigate the possibility and further on the mechanism of obtaining fabrics' tactile properties via samples' visual information. These two tasks are realized progressively through two sensory experiments on a set of flared skirts made of different textile fabrics. From the analytic point of view, a novel approach based on the fusion of rough sets (Pawlak 2007), fuzzy inference (Dubois et al. 1997; Takagi and Sugeno 1985), fuzzy neural network (Jang 1993), and statistical methods (Daniel 1978; Pérez and Barba-Romero 1995) is proposed to explore the relations between different sensory information, so as to realize the visual interpretation of fabric tactile properties.

3.1 Consistency Between Visual and Haptic Perception of Fabric Tactile Properties

3.1.1 Fabric Tactile Evaluation Through Real-Touch and Visual Scenarios

(1) Experimental materials

Eighteen representative fabrics of various tactile properties were selected and made into the two-pieces flared skirts of the same production parameters as our experimental objects. A twenty-five-years-old girl of normal body shape who is 1.68 m tall and weighs 50 kg was invited to be the mannequin in our experiments. A series of multi-angle digital photographs and a video clip were created for each sample skirt representing its static and dynamic performance, respectively.

a. Static visual representation

A digital single-lens reflex (DSLR) camera whose maximum resolution is $5,616 \times 3,744$ pixels was employed to take photographs of each skirt from 8 different directions, according to the angle between the mannequin's front and the position of the lens, 0°, 45°, 90°, 135°, 180°, 225°, 270°, and 315°, respectively.

Table 4 Twenty-one bipolar descriptors

Nm	Descriptor pair	Nm	Descriptor pair	Nm	Descriptor pair
D1	Stiff–pliable	D8	Thin–thick	D15	With ridges–without ridges
D2	Dead–lively	D9	Light–heavy	D16	Bumpy–non-bumpy
D3	Draped–non-draped	D10	Soft–hard (in compression)	D17	Prickly–non-prickly
D4	Crumply–wrinkle-resistant	D11	Non-springy–springy	D18	Fuzzy–non-fuzzy
D5	Non-stretchy–stretchy	D12	Non-full–full	D19	Non-slippery–slippery
D6	Loose–tight	D13	Rough–smooth (overall feeling)	D20	Harsh–soft
D7	Flimsy–firm	D14	Grainy–non-grainy	D21	Warm–cool

b. Dynamic representations

A camcorder whose maximum resolution is 1920×1080 pixels was employed to create video clips for all the sample skirts. In each video clip, the mannequin was required to walk to-and-fro (with pauses) and make various postures according to predefined guidelines, whose aim is to show as comprehensively as possible the static and dynamic performance of the samples.

(2) Evaluation descriptors

In this part of the study, a normalized procedure from brainstorming, screening, adjusting, and final determination has been taken to generate descriptive terminologies for the evaluation of fabric tactile properties of the samples through real-touch and visual scenarios.

Finally, twenty-one pairs of tactile descriptors have been determined and shown in Table 4.

(3) Sensory evaluation

A total of 42 panelists of textile background (30 females and 12 males aged between 23 and 55) including university professors (including lecturers, researchers, or research assistants from two textile colleges), professionals in textile industry (working mainly on fabric design and fashion design), Ph.D., and postgraduate students (working or studying in textile schools) were recruited and organized into three groups (14 panelists per group) to evaluate the 18 samples according to the 21 fabric tactile properties in real-touch, video, and image scenarios using eleven-point (degreed from 0 to 10) evaluation scale.

a. Real-touch scenario

All the samples were conditioned for a minimum of 24 h under the standard atmospheric condition (20 ± 2 °C temperature, $65 \pm 2\%$ relative humidity). Fourteen panelists took part in the real-touch scenario. They were allowed to both see and touch the fabric during the evaluation, which is exactly what they do every day. Before getting started, the panelist would be asked to wash and dry his/her hands with the non-moisturizing soap and paper towel provided. The evaluation should be carried out individually for each panelist.

b. Video scenario

Another fourteen panelists participated in the video assessment scenario, where the video clips of the sample skirts were displayed one by one on a computer screen. During the evaluation, the panelists were free to control the playback of the video clips and make pauses wherever they needed. The panelists were also required to conduct the tests individually.

c. Image scenario

The left fourteen panelists took part in image scenario, in which the multi-angle images of the 18 samples were shown one by one on a computer screen. During the process, the panelists were free to control the display of the images by either changing the displaying order or zooming in/out the image. The evaluations were carried out individually for each panelist.

3.1.2   General Consistency Based on Rough Inclusion and Fuzzy Logic

In this part of the study, an intelligent approach has been proposed to measure the consistency of tactile information obtained through vision with respect to that through real haptic evaluation. This approach is based on a so-called fuzzy inclusion degree defined according to rough sets theory and fuzzy techniques. Compared with the existing statistical methods, our approach is capable of treating relations between variables or attributes from a small number of learning data (Xue et al. 2012).

The framework of this approach is illustrated in Fig. 4. It is constructed upon two major indices, the classification consistency ($CCons*$) based on the concept of fuzzy inclusion degree, and the ranking consistency ($RCons$) obtained from the nonparametric ranking coefficient (Kendall's τ), respectively. Then, in order to generate a criterion to measure the general consistency (denoted by $GCons$) between different datasets in the condition that this criterion should be both robust to noise and easy for interpretation, a fuzzy inference system is developed to integrate the previous two indices, $CCons*$ and $RCons$, respectively.

(1) Classification consistency ($CCons*$)

The formalization of the present problem is given below. Let $U = \{e_1, e_2, ..., e_{18}\}$ be the set of samples. The corresponding evaluation scores for any specific pair of tactile descriptors such as 'stiff–pliable' have been obtained from the visual (either image or video) and real-touch perceptions on all the samples, denoted as $C = (c(e_1)$ ... $c(e_{18}))^T$ and $D = (d(e_1) ... d(e_{18}))^T$, respectively. All the evaluation scores $c(e_i)$'s

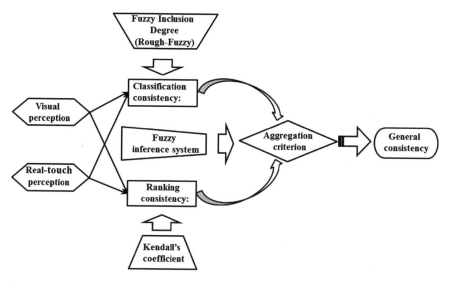

Fig. 4 General framework of the proposed consistency measure

and $d(e_i)$'s vary between 0 and 10. In the following discussion, the visual perception C is taken as condition variable and the real-touch perception D as decision variable. According to rough sets philosophy, the knowledge acquisition is in fact a process of knowledge classification. Different knowledge would generate different partitions of data. From the previous two vectors of C and D, we obtain two partitions for the visual and real-touch results, i.e., $U/C = \{X_0, X_1, ..., X_{10}\}$ and $U/D = \{Y_0, Y_1, ..., Y_{10}\}$. $X_i \in U/C$ ($i = 0, 1, ..., 10$) (or $Y_j \in U/D$ ($j = 0, 1, ..., 10$)) is the class of samples in which the evaluation scores are all i (or all j). In practice, some X_i (or Y_j) can be empty if its index i (or j) does not exist in the evaluation scores of all the samples e_k's.

The classification consistency is proposed based on the idea of inclusion degree from rough sets theory (Pawlak 1998). According to rough sets theory, knowledge acquisition is in fact a process of knowledge classification. Different knowledge would generate different partitions of data. In this sense, as the first important index of the consistency measure, the aim of this part is to compute the extent to which the classification of the visual data (tactile perceptions obtained from video and image scenarios) is consistent with that of the real-touch data (tactile perceptions from real-touch scenario) during the evaluation of fabric tactile properties.

a. Inclusion degree

Let $S = (U, C \cup D)$ be a complete decision table, U the collection of experimental samples, $X \in U/C$ an equivalence class representing the visual perceptions of the samples on a specific tactile property and $U/D = \{[e]_D : e \in U\}$ representing the corresponding real-touch perceptions of the samples. For any sample $e_k \in U$, the inclusion degree of X_i with respect to $[e_j]_D$ is denoted by:

$$Inc\left([e_k]_D / X_i\right) = \frac{|X_i \cap [e_k]_D|}{|X_i|}, \quad (i = 0, 1, \ldots, 10; \ k = 1, 2, \ldots, 18) \quad (11)$$

where $0 \leq Inc\left([e_k]_D / X_i\right) \leq 1$.

$[e_k]_D$ is the set of samples that are classified into the same group with e_k according to the decision attribute.

In fact, this formulation of inclusion degree is in agreement with that of the rough membership function of e_k in X_i, i.e., $\delta_{X_i}(e_k) = \frac{|X_i \cap [e_k]_D|}{|X_i|}$ (Pawlak 1982), defined according to rough sets theory. It is evident that, if $Inc\left([e_k]_D / X_i\right) = 1$, then X_i can be said to be consistent with respect to $[e_k]_D$, or one has $X_i \subseteq [e_k]_D$ (which is called a complete inclusion).

b.   Fuzzy inclusion degree

In practice, an inclusion degree based on the crisp partition of samples might lead to serious information loss. For any specific class, it does not make difference between samples close to it and those far from it. For any sample e_k not belonging to a class X_i, we have its membership value equaling zero, indicating that its adhesion to this equivalence class is regarded as null. Evidently, in the definition of the inclusion degree, it is more reasonable to consider the degrees of adhesion of the samples to the equivalence classes so that samples close to a class are more important than those far from it. According to this idea, we modify the previous inclusion degree using the concept of fuzzy partition.

The following illustrates the modified inclusion degree $FInc\left([e_k]_D / X_i\right)$

$$FInc\left([e_k]_D / X_i\right) = \sum_{l=1}^{18} \min\left(\mu_{X_i}(e_l), \mu_{[e_k]_D}(e_l)\right) / \sum_{l=1}^{18} \mu_{X_i}(e_l) \quad (12)$$

in which X_i denotes the ith condition set and $[e_k]_D \subset U/D = \{Y_0, Y_1, \ldots, Y_{10}\}$ is the decision set where the sample e_k belongs. The classes X_i's and Y_j's constitute two fuzzy partitions (sets) for the attributes C and D, and they are characterized by the fuzzy membership functions $\mu_{X_i}(e)$ and $\mu_{Y_j}(e)$ defined as:

$$\mu_{X_i}(e) = 1 - h\,|i - c(e)| \quad (13)$$
$$\mu_{Y_j}(e) = 1 - h\,|j - d(e)| \quad (14)$$

They are triangular functions centered on i and j, respectively. h is the coefficient controlling the sensitivity of these functions. In the current study, we assign 0.2 to h as a general case.

Notably, (1) the decision set where a sample e_k might belong is determined according to the maximum membership principle, i.e., e_k is believed to belong to a decision set Y_j when the corresponding fuzzy membership degree $\mu_{Y_j}(e_k)$ reaches the highest among all the decision sets; (2) given the present problem, any $\mu_{X_i}(e)$ or $\mu_{Y_j}(e)$ whose value is negative would be taken as zero for further computation. Figure 5 depicts the membership function of the descriptor pair 'stiff–pliable' as an example.

c.   Classification consistency based on fuzzy inclusion

Fig. 5 Membership function of 'stiff–pliable'

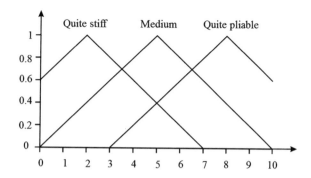

The classification consistency of X_i with respect to D has been modified as:

$$Cons^*(X_i, D) = 1 - \frac{4}{\sum_{i=0}^{10} \sum_{k=1}^{18} \mu_{X_i}(e_k)} \sum_{k=1}^{18} FInc\left([e_k]_D / X_i\right)\left(1 - FInc\left([e_k]_D / X_i\right)\right)$$

(15)

where $FInc\left([e_k]_D / X_i\right)$ is the fuzzy inclusion degree of X_i into $[e_k]_D$ for sample e_k. We have $0 < Cons^*(X_i, D) < 1$. This is due to the following facts: (1) for each sample e_k, there always exists a Y_j whose value j is the closest to $d(e_k)$ (the difference is smaller than 0.5), i.e., $\mu_{Y_j}(e_k) \geq 0.9$; (2) $0 \leq FInc\left([e_k]_D / X_i\right)\left(1 - FInc\left([e_k]_D / X_i\right)\right) \leq \frac{1}{4}$ holds for any X_i and Y_j.

Hitherto, the final classification consistency of the condition attribute C with respect to the decision attribute D can be constituted as:

$$CCons^*(C, D) = \sum_{i=0}^{10} \frac{\sum_{k=1}^{18} \mu_{X_i}(e_k)}{\sum_{i=0}^{10} \sum_{k=1}^{18} \mu_{X_i}(e_k)} Cons^*(X_i, D)$$

(16)

where $FInc\left([e_k]_D / X_i\right)$ is the fuzzy inclusion degree of X_i into $[e_k]_D$ for sample e_k.

(2) Ranking consistency ($RCons^*$)

In an information system, the element's different positions in the two sets may lead to big differences between the knowledge to be represented, respectively. Therefore, it is significant to include another index to measure the ordinal difference between the data obtained from different sensory modalities. For this purpose, Kendall's rank coefficient is introduced in our study.

As a nonparametric measure of rank correlation, Kendall's τ depends upon the number of inversions of pairs of objects which would be needed to transform one rank order into the other. Still take the first experiment as an example, for the sample set $U = \{e_1, e_2, \ldots, e_{18}\}$, $C = (c(e_1) \ldots c(e_{18}))^T$ and $D = (d(e_1) \ldots d(e_{18}))^T$ are observations corresponding to the visual evaluation (i.e., for the video or image scenarios) and the real-touch evaluation (i.e., for the real scenario), respectively. Kendall's rank correlation coefficient τ or $RCons(C, D)$ is computed from:

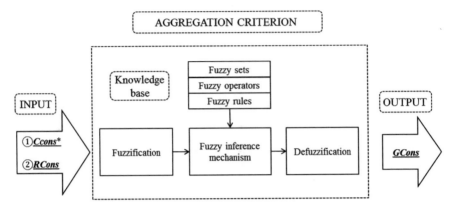

Fig. 6 *AC* constituted based on a fuzzy inference system

$$RCons(C, D) = \tau(C, D) = 1 - \frac{2d_\Delta(C, D)}{n(n-1)}$$

$$-1 \le \tau(C, D) \le 1 \tag{17}$$

where n is the number of samples. $d_\Delta(C, D)$ denotes the symmetric difference between two attribute sets C and D, and it is obtained from the following formula:

$$d_\Delta(C, D) = (number\ of\ concordant\ pairs) - (number\ of\ discordant\ pairs) \tag{18}$$

Notably, any pair of observations $(c\,(e_i)\,, d\,(e_i))$ and $(c\,(e_j)\,, d\,(e_j))$ are considered to be concordant if the ranks for both elements agree: i.e., if both $c\,(e_i) > c\,(e_j)$ and $d\,(e_i) > d\,(e_j)$ or if both $c\,(e_i) < c\,(e_j)$ and $d\,(e_i) < d\,(e_j)$. On the other hand, they are considered to be discordant, if $c\,(e_i) > c\,(e_j)$ and $d\,(e_i) < d\,(e_j)$ or if $c\,(e_i) < c\,(e_j)$ and $d\,(e_i) > d\,(e_j)$. But if $c\,(e_i) = c\,(e_j)$ or $d\,(e_i) = d\,(e_j)$, then the pair is neither concordant nor discordant.

(3) General consistency measure (*GCons*)

Here, an aggregation criterion (*AC*) was proposed to integrate the previous two indices, *CCons** and *RCons*, so as to constitute a general consistency measure (*GCons*) to investigate the overall consistency between the two sensory modalities. This criterion should be both robust to noise and easy for knowledge interpretation. For this purpose, a fuzzy inference system (Wang et al. 2009) is designed and illustrated in Fig. 6. This system consists of three major parts responsible for the fuzzification of input data, fuzzy operation based on fuzzy rules, and defuzzification to produce output data, respectively.

Fuzzy rules are crucial for building a fuzzy inference system. In the current study, discussions were carried out among a panel of 6 experts to design a series of fuzzy rules. *VS, S, M, L, VL* denote the linguistic values of '*Very small*,' '*Small*,' '*Medium*,'

Fig. 7 Fuzzy membership function for input/output variables

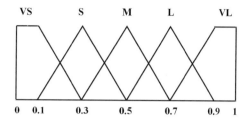

'*Large,*' and '*Very Large,*' respectively. For example, a fuzzy rule can be described as the following linguistic expression:

IF CCons* is *medium (M)* and *RCons* is *small (S)*, **THEN** GCons is *small (S)*.

Each linguistic expression is equivalent to a numerical value ranged from 0 to 1 according to the fuzzy membership function defined for each input and output variable during the process of fuzzy inference. The membership function is illustrated in Fig. 7. It is a commonly used function uniformly distributed on [0, 1].

After applying the *AC*, a general consistency measure (*GCons*) was constituted to investigate the extent to which the tactile properties can be transmitted through different visual representations of a textile product, with both the classification consistency and the distribution similarity taken into consideration.

(4)   Modeling results and discussion

In this part of the study, three groups of panelists were asked to assess the (same set of) tactile properties of the samples in real-touch (or haptic), video, and image scenarios, the aim of which is to measure the consistency of the tactile information perceived through vision with respect to that perceived by real-touch.

GCons results are computed and illustrated as the so-called Perceptual lines for both video and image observations with respect to the reference (i.e., the real-touch observation) in Fig. 8. The solid line represents the video performance, while the dashed line represents the image performance. This figure reveals that, out of the overall 21 descriptors, 19 (over 90%) have their consistency values higher than 0.6 and 13 (about 62%) higher than 0.7 in both video and image scenarios, which indicates that an important part of the tactile properties can be well interpreted through visual representations of the apparel products.

On most of the descriptors, the shape of the solid line (video data) is more stable than the dashed line with fewer fluctuations. Besides, in video scenarios, there are 18 out of 21 descriptors whose *GCons* values are higher than 0.7 (i.e., over 85%). All these indicate that the performance of the panelists in video scenarios is generally better with more stability and higher accuracy as compared with that in the image scenarios. It is reasonable since compared with static photographs captured from a limited number of angles video clips can record information about a sample from every possible angle and in a continuous way just as it is viewed in the real experience.

Besides, both the solid line and dashed line shape generally more and larger fluctuations than for the descriptors on the right part of the figure. This indicates

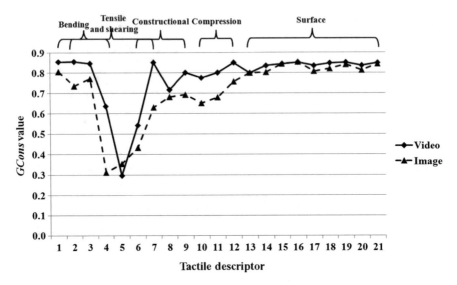

Fig. 8 *Gcons* values on tactile descriptors for visual and haptic scenarios

that the panelists in either video or image scenarios tended to encounter difficulties in assessing the descriptors concerning samples' mechanical properties. Actually, the assessment of mechanical properties (e.g., stretchiness, tightness and firmness) depends on direct handle of the fabric, such as stretching, grasping, bending. Thus when touch is deprived, the accuracy of the assessment of these properties will be challenged.

On the other hand, satisfactory performance is found on the assessment of fabrics' surface properties which graphically are positioned on the right part of the figure. Both the two perceptual lines are comparatively stable in shape and stay above the satisfactory level of 0.8. This indicates that although the panelists were not allowed to really touch the fabric, they are still able to perceive most of the fabrics' surface properties with quite a certainty.

To be brief, the first sensory experiment tells us that, for the samples under discussion, (1) most fabric tactile properties could be well perceived through samples' visual (image and video) displays; (2) panelists participated in video scenarios has more stable and correct performance than those in image scenarios. (Videos transmit more reliable and more comprehensive tactile information than images.); (3) Fabrics' surface properties could be well perceived through both static and dynamic display of the samples, while mechanical properties are comparatively more difficult to be accurately perceived through the present way of visual representation.

Table 5 Visual features of the samples

Position	Nm	Feature		Position	Nm	Feature	
Waistline	E1	Outline of pleats	Clear–fuzzy	Luster	E17	Light intensity	Weak–strong
	…	…	…	Silhouette	E18	Fitness	Unfit–fit
Abdomen and hip	E3	Fitness to body shape	Unfit–fit		…	…	…
	…	…	…	Dynamic	E22	Balance	Badly–well
Lower part of skirt	E7	Expending extent from below the hip	Not expand-ing–ex-panding		…	…	…
	…	…	…	Fabric	E29	Rough–smooth (overall feeling)	
Color	E14	Brightness	Dark–light		…	…	
	…	…	…				

3.2 Visual Interpretation of Fabric Tactile Properties

3.2.1 Evaluation of Samples' Visual Features

On the basis of Experiment 1, a second experiment (Experiment 2) was carried out by a new group of panelists to assess the visual features of the eighteen sample skirts.

(1) Evaluation descriptors

This part of the study is to generate a list of descriptors to cover as comprehensively as possible the visual features of the sample skirts. These descriptors should be composed of two aspects revealing, respectively, the aesthetic appearance about the skirts and the visual characteristics about the fabrics.

After normalized procedures similar to those for the tactile evaluation, twenty-eight descriptors covering both the static and dynamic attributes of the sample skirts have been determined. Besides, from the generated tactile descriptors, eight descriptors concerning fabric surface properties (from D13 to D20) and one concerning fabric thickness (D8) have been categorized as visible fabric features. Therefore, the overall visual descriptors consist of thirty-seven

In this study, the visual descriptors involve both the skirt appearance features ($n = 28$) and the above-discussed visible fabric features ($n = 9$). Therefore, there are in total thirty-seven visual descriptors some of which are listed in Table 5.

(2)   Sensory evaluation

Five professionals were recruited from the apparel industry to evaluate skirts' visual features. All these panelists have profound experience in evaluating the appearance of apparel products according to standard criteria.

According to the results obtained from Experiment 1, panelists in video scenarios had better performance than those in image scenarios. Thus in Experiment 2, the panelists were required to watch only the video clips of the samples. Besides, Experiment 1 tells us that the available video display of the samples might not be adequate for well illustrating some mechanical properties (e.g., tensile, shearing, bending properties, etc.). Therefore, in Experiment 2, apart from the same video clip used in Experiment 1, for each sample skirt, another video clip is created to show more straightforward the mechanical properties by recording a subject performing stretching, bending, and grasping gestures (etc.) on the sample.

During the evaluation, the panelists were free to control the playback of the video clips and make pauses wherever they needed. An eleven-point semantic scale has been used in the evaluation of visual features in Experiment 2.

3.2.2   Visuo-haptic Predictive Modeling

Based on the resolution of the first problem, the second problem of the study is to further investigate the quantitative mechanism of visual interpretation of each fabric tactile property. To realize this, from the mathematical point of view, two tasks should be accomplished. The first task is to extract for each tactile property the most relevant visual features called the principal visual features (*PRF*s). On this basis, the second or the final task is to develop a mathematical model so as to quantify the previously obtained multiple (visual features)-to-single (real-touch property) relation for each tactile property.

(1)   Feature selection

The consistency measure applied to the first problem deals with single-to-single relation between the tactile perception obtained from vision and that from real-touch. In this part of the study, a multiple-to-single relation should be explored between any tactile property and its principal visual features. The new consistency measure is a further development of the previous one.

Since the previous consistency measure is composed of two indices, classification consistency and ranking consistency, modifications should be applied to these two indices, respectively, to constitute the new measure.

With respect to the classification consistency, since for a certain sample, its sensory score on each visual variable (or feature) is represented by a fuzzy set, the aggregation of different sets is realized by the operation of fuzzy union (Klir et al. 1995).

According to the principles of fuzzy set operation (Dubois et al. 1997), in the current problem, the union is defined as the smallest fuzzy set containing all the l fuzzy sets to be aggregated. Let fuzzy set A be the fuzzy union result written as

$A(e_k) = E_1(e_k) \cup E_2(e_k) \cup \cdots \cup E_l(e_k)$. Its membership function is constituted as follows:

$$f_A(e_k) = \max\left\{f_{E_1}(e_k), f_{E_2}(e_k), \ldots, f_{E_l}(e_k)\right\}, \quad e_k \in U \tag{19}$$

In this manner, a fuzzy matrix containing the results for all the samples will be obtained as the new condition table. According to this idea, the classification consistency of the condition attribute C with respect to the decision attribute D has been finally constituted as:

$$FCons(C, D) = \sum_{i=0}^{t} \frac{\sum_{k=1}^{m} \mu'_{X_i}(e_k)}{\sum_{i=0}^{t} \sum_{k=1}^{m} \mu'_{X_i}(e_k)} AFCons(X_i, D) \ (m = 18, \ t = 10) \tag{20}$$

where $AFInc([e_k]_D / X_i)$ is the aggregated fuzzy inclusion degree of X_i into $[e_k]_D$ for sample e_k.

On the other hand, a statistical method named Borda's rule (Pérez et al. 1995) is applied to aggregate for each sample its ordinal position with respect to different visual variables. A vector containing the aggregated rankings of all the samples is obtained in this manner. The method of Borda's rule is realized by Microsoft Excel Statistics (Salkind 2010).

To examine the ranking consistency between the aggregated visual features and the corresponding tactile properties is to compute Kendall's ranking coefficient.

Finally, a fuzzy inference system is developed to integrate the two modified indices to constitute the new general consistency measure denoted as **Acons** which is capable of exploring the extent to which a certain fabric tactile property could be interpreted through a set of relevant visual features.

By applying the modified consistency measure (**Acons**), for each tactile property, the visual features whose cooperative consistency value is highest will be extracted as the most relevant.

(2) Quantitative prediction

This part of the study is aimed to set up a mathematical model to quantify the multiple-to-single relations obtained from Task 1. Given that we have a relatively small number of samples, and there exists much uncertainty and imprecision in sensory data, an adaptive-network-based fuzzy inference system (ANFIS) (Jang 1993; Behera et al. 2012; Rwawiire et al. 2014) is employed to realize the modeling since it well absorbs the advantages of neural network and fuzzy inference system being capable of both pattern learning and producing interpretable results.

By employing ANFIS, the complicated relations between each tactile property and its principal visual features is modeled with effectiveness and interpretability. In the current study, the ANFIS model is realized using MATLAB 2014b.

(3) Modeling results and discussion

As mentioned previously, among the 21 tactile properties, there are 11 who are invisible (including D1, D2, D4, D5, D6, D7, D9, D10, D11, D12, and D21), while the left 10 descriptors are visible.

The aim of this part is to extract for each invisible tactile property a series of principal visual features. A systematic method is proposed to realize this in two steps based on, respectively, the single-to-single consistency and the multiple-to-single consistency.

(1) Selection of relevant visual features

For each visual feature, its *GCons* value (single-to-single consistency) is computed with respect to each tactile property. For each tactile property, we obtain a set of *GCons* values corresponding, respectively, to the impacts of 37 visual features. For each tactile property, its highest **GCons** value calculated from the 37 visual features is named the 'Impact level.' Upon the observation on all the 11 tactile properties, it is determined that for a certain tactile property, those visual features whose **GCons** values are higher than the 70% of the impact level are considered as the relevant visual features (*RF*s).

Taking the tactile descriptor pair 'Dead–lively' (D2) as an example, Fig. 9 shows its *GCons* values ranked in descending order. For any tactile property, on the sequence of its **RF**'s, if a certain visual feature is found to be the first one whose decreasing rate exceeds the average decreasing rate, those visual features before (and containing) it are considered as the **MRF**'s. As an example, for the descriptor pair 'Stiff–pliable' (D1), its **MRF**'s include 'overall outline' (E21), 'Fitness to body shape' (E3), 'Drape' (E20), 'Expending extent from below the hip' (E7), 'Clingingness' (E24), 'Ethereality' (E25), 'Wave flowability' (E26), and 'Rigidity of pleats' (E2).

(2) Selection of principal visual features

In fact, to understand the impact of several visual features on the perception of one tactile property, the multiple-to-single relations must be studied, since the impact of any visual feature is not independent. The aim of this section is to extract on the basis of the previous exploration the visual features that have the best cooperative impact on each tactile property or the so-called principal visual features (**PRF**s).

According to the results obtained previously, each tactile property corresponds to a sequence of **MRF**'s whose impacts (or **GCons** values) are ranked in descending order. On this sequence, the consistency is measured of the first two visual features with respect to the tactile property. The obtained value is compared with the impact level (defined previously as the highest **GCons** value that could be obtained from all the visual features) of the tactile property. If it is smaller, then the first visual feature is regarded as the single PRF. If it exceeds the impact level, these two features are both taken as the **PRF**'s and then the third feature on the sequence will be involved

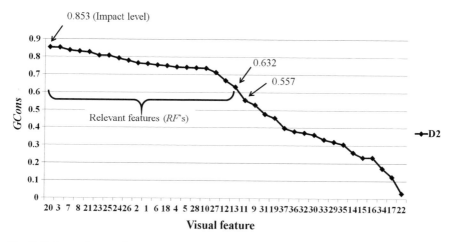

Fig. 9 Relevant features for Dead–lively (D2)

to compute the 3-to-1 consistency, and the result of which will be compared with the previously obtained two-to-one consistency value to see if the third feature largely contribute to the interpretation of the tactile property of interest. The impact of the rest of the features on the sequence is examined in this manner, and finally, the group of visual features that has the highest cooperative consistency will be considered as the **PRF's**.

Figure 10 shows the example of descriptor D2 ('Dead–lively'). According to our criterion, the cooperative consistency of the first three visual features on the sequence reaches 0.871 which is the highest among all the combinations. So the drape of the skirt (E20), the fitness at abdomen and hip (E3), and the expanding degree of the lower part of the skirt (E7) are taken as the three most relevant features to reveal the samples' liveliness. In this manner, for each tactile property, a series of principal visual features (**PRF**s) are extracted.

(3) Predictive modeling

a. Modeling

As mentioned previously, the fuzzy neural network ANFIS has been used to model the multiple-to-single relationship between each tactile property (Output variable) and its principal visual features (Input variables). In addition to the eighteen samples taken as the training data, another three samples of diverse tactile properties were prepared as the testing data for the model validation. As for the configuration, bell-shaped membership function is used in the study. In order to ensure the predictive precision of the model and at the same time reduce the complexity of the system, it is determined that for the tactile properties who have more than two (including two)

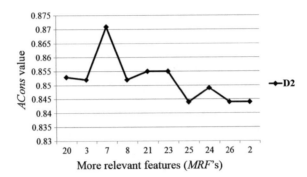

Fig. 10 Cooperative impacts of the *MRF*'s for D2

principal visual features, three membership functions are assigned; for those who have only two *PRF*'s, five membership functions are assigned.

According to the ANFIS structure, in each fuzzy rule, the final output can be expressed as a linear combination of the consequent parameters (Ozturk et al. 2008). For better understanding, suppose we have a fuzzy if-then rule as a general case as follows:

$$\text{If } x \text{ is } A \text{ and } y \text{ is } B \text{ then } f = px + qy + r$$

where f is the output variable of the ANFIS, x and y are the input variables. A and B are the fuzzy sets representing the ranges of value for different input variables. p, q, and r are the so-called consequent parameters.

The hybrid learning algorithm of the ANFIS combines the gradient method with the least squares method to update (or train) the parameters (Cárdenas et al. 2012).

In our study, taking descriptor D2 as an example, after parametric training, a total of 27 ($n = 3^3$) if-then rules have been extracted to describe the mathematical relation between the output variable (i.e., D2) and the input variables (i.e., the *PRF*'s, E20, E3, and E7) of the predictive model (ANFIS). The following shows one of the extracted fuzzy rules for D2.

If (E20 is low) and (E3 is low) and (E7 is medium),

then D2 $= 0.6532 * E20 + 0.3938 * E3 + 0.6362 * E7 + 0.1997$

According to this rule, if we have a sample with E20$=3.1$ (low), E3$=2.3$ (low) and E7$=6.2$ (medium) as evaluated by the assessors, the evaluation score of the tactile property D2 will be estimated to be 7.1 which corresponds to the semantic value 'Fairly lively' according to the differential scale.

b. Verification

For each sample, a set of predictive tactile properties are obtained by applying our ANFIS models to the evaluation values of the principal visual features. Figure 11 shows the comparison between the experimental and the predictive values of the tactile properties for the three samples, respectively. It is obvious from the three graphs that for most of the tactile properties, the predictive results are quite close to the experimental ones with the biggest difference not exceeding 1, which could be considered quite satisfactory since for sensory science the conventional concept of accuracy is of little significance.

Therefore, it is concluded that, as far as it is concerned in this study, the developed predictive system is capable of well-estimating fabrics' tactile properties from the perceived visual features of the samples. So far, the visual mechanism of fabrics' tactile properties is unveiled with success.

4   General Conclusion

Human perception can be the vaguest since it is often difficult to quantify and control. But at the same time, it can be considered as the most accurate due to the fact that it transmits directly and truthfully what we feel about the surrounding world. Sensory evaluation that emerged during the 1970s provides a good opportunity to collect and analyze the ambiguous objects and complex relations involved in human knowledge in a scientific and standardized manner. Among the various data analysis methods that have been used in sensory science, artificial intelligent tools are showing their overwhelming advantages in recent years due to their high capacity in dealing with the uncertainty and precision born with human data and the comparatively lower request on data distribution and sample size. Fabric tactile evaluation is a representative application of sensory evaluation methods to the study of textile products. In this chapter, the relations between perceptions of fabric tactile properties acquired from different sensory modalities (i.e., visual and haptic perceptions) and different perceptual levels (fabric hand evaluation and fabric total preference) have been investigated using artificial intelligent tools such as fuzzy inference, neural networks, and rough inclusion within the framework of sensory evaluation methodologies, in order to build up direct and effective communication with the tactile properties of textile products from a multisensory perspective.

Nowadays, caring about human has become the recipe for stakeholders in both industrial and fashion designs. Any product being a washing machine or an evening dress should pay enough attention to customers' need, or it has a good chance to fail in the increasingly fierce market competition. The recent emergency of wearable technology and the various products and services driven by artificial intelligence provide a vital illustration to the human-oriented design concept.

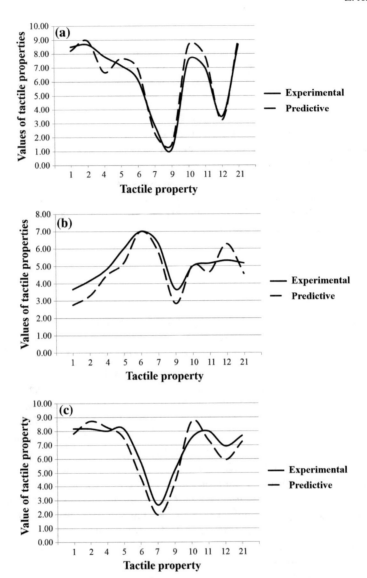

Fig. 11 Comparison between experimental and predictive tactile properties for **a** testing sample 1; **b** testing sample 2; **c** testing sample 3

Sensory evaluation of textile products is an important branch of human-centered fashion design which deals directly with one's physical and psychological comfort and further on affects his (or her) experience about well-being. In this sense, various intelligent computing technologies that are believed to be more capable of handling human data will definitely be an outstanding tool in this domain.

Acknowledgements This work was financially supported by the National Natural Science Foundation of China (No. 61503154) and the Fundamental Research Funds for the Central Universities (No. JUSRP11503).

References

Ali SI, Begum S (1994) Fabric softeners and softness perception. Ergonomics 37:801–806

Behera B, Guruprasad R (2012) Predicting bending rigidity of woven fabrics using adaptive neuro-fuzzy inference system (ANFIS). J Text Inst 103:1205–1212

Bernard W, Juan CA (2005) 'Cognitive' memory. In: Proceeding of international joint conference on neural networks, Montreal, Canada, 31 July–4 Aug 2005, pp 3296–3299

Cárdenas JJ, Romeral L, Garcia A, Andrade F (2012) Load forecasting framework of electricity consumptions for an Intelligent Energy Management System in the user-side. Expert Syst Appl 39:5557–5565

Daniel WW (1978) Applied nonparametric statistics. Houghton Mifflin

Davis L (1991) Handbook of genetic algorithms

Dijksterhuis GB (2008) Multivariate data analysis in sensory and consumer science. Wiley

Dubois D, Prade H, Yager RR (1997) Fuzzy information engineering: a guided tour of applications. Wiley

Fausett L (1994) Fundamentals of neural networks: architectures, algorithms, and applications

Fernandes A, Albuquerque PB (2008) Tactile perceptual dimensions: a study with light-weight wool fabrics. In: International conference on human haptic sensing and touch enabled computer applications. Springer, pp 337–342

Giboreau A, Navarro S, Faye P, Dumortier J (2001) Sensory evaluation of automotive fabrics: the contribution of categorization tasks and non verbal information to set-up a descriptive method of tactile properties. Food Qual Prefer 12:311–322

Goldberg D (1989) Genetic algorithms in search, optimization and learning. Addison-Weseley

Härdle W, Simar L (2010) Applied multivariate statistical analysis: second edition. Clustering, distance methods, and ordination

Holliins M, Faldowski R, Rao S, Young F (1993) Perceptual dimensions of tactile surface texture: a multidimensional scaling analysis. Percept Psychophys 54:697–705

Howorth WS, Oliver PH (1958) The application of multiple factor analysis to the assessment of fabric handle. J Text Inst Trans 49:T540–T553

ISO8586 1993. Assessors for sensory analysis, Part 1: Guide to the selection, training and monitoring of selected assessors. International Standard ISO-8586–1993 (F)

ISO11035 1995. Sensory analysis—identification and selection of descriptors for establishing a sensory profile by a multidimensional approach. International Standard ISO-11035-1995 (F)

Jang J-S (1993) ANFIS: adaptive-network-based fuzzy inference system. IEEE Trans Syst Man Cybern 23:665–685

Jeguirim SEG, Dhouib AB, Sahnoun M, Cheikhrouhou M, Njeugna N, Schacher L, Adolphe D (2010) The tactile sensory evaluation of knitted fabrics: effect of some finishing treatments. J Sens Stud 25:201–215

Jolliffe IT (2010) Principal component analysis, vol 87. Springer, Berlin, pp 41–64

Klir G, Yuan B (1995) Fuzzy sets and fuzzy logic. Prentice Hall, New Jersey

le Dien S, PAG S J (2003) Hierarchical multiple factor analysis: application to the comparison of sensory profiles. Food Qual Prefer 14:397–403

McGregor BA, Naebe M, Wang H, Tester D, Rowe J (2015) Relationships between wearer assessment and the instrumental measurement of the handle and prickle of knitted wool fabrics. Text Res J 85:1140–1152

Ozturk A, Arslan A, Hardalac F (2008) Comparison of neuro-fuzzy systems for classification of transcranial Doppler signals with their chaotic invariant measures. Expert Syst Appl 34:1044–1055

Pérez J, Barba-Romero S (1995) Three practical criteria of comparison among ordinal preference aggregating rules. Eur J Oper Res 85:473–487

Park S-W, Hwang Y-G, Kang B-C, Yeo S-W (2000) Applying fuzzy logic and neural networks to total hand evaluation of knitted fabrics. Text Res J 70:675–681

Pawlak Z (1982) Rough sets. Int J Parallel Prog 11:341–356

Pawlak Z (1998) Rough set theory and its applications to data analysis. Cybern Syst 29:661–688

Pawlak Z, Skowron A (2007) Rudiments of rough sets. Inf Sci 177:3–27

Philippe F, Schacher L, Adolphe DC, Dacremont C (2004) Tactile feeling: sensory analysis applied to textile goods. Text Res J 74:1066–1072

Picard D, Dacremont C, Valentin D, Giboreau A (2003) Perceptual dimensions of tactile textures. Acta Psychol (Amst) 114:165–184

Ruan D (1997) Intelligent hybrid systems: fuzzy logic, neural networks, and genetic algorithms. Kluwer Academic Publishers

Rwawiire S, Kasedde A, Nibikora I, Wandera G (2014) Prediction of polyester/cotton ring spun yarn unevenness using adaptive neuro Fuzzy inference system. J Text Appar Technol Manag 8

Salkind NJ (2010) Excel statistics. Sage Publications Ltd

Shinomoto S (1987) A cognitive and associative memory. Biol Cybern 57:197–206

Stone H, Bleibaum R, Thomas HA (2012) Sensory evaluation practices. Academic press

Suelar V, Okur A (2007) Sensory evaluation methods for tactile properties of fabrics. J Sens Stud 22:1–16

Sugeno M, Yasukawa T (1993) A fuzzy-logic-based approach to qualitative modeling. IEEE Trans Fuzzy Syst 1

Takagi T, Sugeno M (1985) Fuzzy identification of systems and its applications to modeling and control. IEEE Trans Syst Man Cybern 116–132

Vogel DD (2005) A neural network model of memory and higher cognitive functions. Int J Psychophysiol 55:3–21

Wang X, Ruan D, Kerre EE (2009) Mathematics of fuzziness—basic issues. Springer Science & Business Media

Wang Y, Li S, Mao J (2001) Computer image processing and recognition technology. Higher Education Press, Beijing

Weisberg S (2005) Applied linear regression. Wiley

Xue Z, Zeng X, Koehl L, Chen Y (2012) Visual interpretation of fabric tactile properties using fuzzy inclusion degree and ANFIS. In: Uncertainty modeling in knowledge engineering and decision making. World Scientific

Yager RR (1977) Multiple objective decision-making using fuzzy sets. Int J Man Mach Stud 9:375–382

Zadeh LA (1965) Fuzzy sets. Inf Control 8:338–353

Zeng X, Da R, Koehl L (2008) Intelligent sensory evaluation: concepts, implementations, and applications. Math Comput Simul 77:443–452

Zeng XY, Koehl L (2003) Representation of the subjective evaluation of the fabric hand using fuzzy techniques. Int J Intell Syst 18:355–366

Zimmermann H-J (2011) Fuzzy set theory—and its applications. Springer Science & Business Media

Evaluation of Fashion Design Using Artificial Intelligence Tools

Yan Hong, Xianyi Zeng, Pascal Brunixaux and Yan Chen

Abstract Fashion design effect evaluation emerges as a new task in automatic fashion product design and development. In this chapter, we give an example of how to evaluate fashion design effect using artificial intelligence tools. The proposed evaluation method is applied to a 3D-to-2D garment design process-based garment design case aiming at the physically disabled people with scoliosis (PDPS). A pattern modification model is proposed in order to satisfy the requirement of design automation. The proposed modification model is based on the evaluation of the 3D virtual try-on result using the professional knowledge of the designers. Research results indicate that 3D-to-2D garment design can effectively realize personalized design for atypical morphology. Also, using sensory evaluation, professional design knowledge of the designers can be fully applied to control the 2D technical space of garment design based on the perception of 3D garment virtual try-on result.

1 Introduction

Artificial intelligence has shown its efficiency in fashion design, such as fashion design effect prediction (Chen et al. 2015), fashion style recommendation (Wang et al. 2015), and color recommendation (Hong et al. 2018). However, evaluation of fashion design has not been investigated in previous studies. In this chapter, a fashion design effect evaluation procedure performed in a virtual environment has been introduced, which can be regarded as a demonstration of the application of artificial intelligence in fashion effect evaluation.

Y. Hong (✉) · X. Zeng · P. Brunixaux
GEMTEX, ENSAIT, 2 allée Louise et Victor Champier, 59056 Roubaix Cedex 1, France
e-mail: yan.hong@ensait.fr

Y. Hong · X. Zeng · P. Brunixaux
Université de Lille, 59655 Lille, France

Y. Hong · Y. Chen
College of Textile and Clothing Engineering, Soochow University, Suzhou 215021, China

© Springer Nature Singapore Pte Ltd. 2018
S. Thomassey and X. Zeng (eds.), *Artificial Intelligence for Fashion Industry in the Big Data Era*, Springer Series in Fashion Business,
https://doi.org/10.1007/978-981-13-0080-6_12

245

With the development of 3D laser body scan and surface reconstruction technology, designing a garment directly on a 3D scanned body using virtual reality and 3D CAD technology emerges as a novel design solution for fashion industry. Currently, the 3D method consists of several design phases, which include 3D scanning, digitalized 3D body modeling, 3D garment generation, interactive 3D garment surface modeling and pattern design. Compared with traditional design method such as traditional 2D-to-3D method, virtual reality-based 3D-to-2D design method is proved to be able to effectively satisfy the personalized garment design for atypical morphology. However, 3D garment simulation is rather time costing and requires of high-level technical support of the software operation. What is more, due to the diversity of garment style, it is difficult to define a systematic design rule for designing ready-to-wear products of all kind of styles using 3D-to-2D design method.

To solve this problem, in our previous study, we developed an interactive virtual try-on-based 3D garment block design for physically disabled people of scoliosis type (PDPS) (Hong et al. 2016a). A personalized garment block is designed for PDPS with controlled wearing ease (Hong et al. 2016a). Using this block, 2D garment patterns of the desired ready-to-wear products can be extended from the proposed garment block pattern (Hong et al. 2016a). In this paper, based on previous study, we introduced an application case of designing a ready-to-wear garment using the garment block design from 3D-to-2D design method in order to validate the applicability of 3D-to-2D design method (Hong et al. 2018). Using personalized garment block, all kinds of garment style can be realized very fast in 2D environment, which can be easily controlled by designers compared with designing in 3D environment.

In this research, a perception-based pattern modification model based on the concept of artificial intelligence is proposed on the basis of a collaborative design process. Perceptual data of consumers on the desired product is extracted and analyzed by the knowledge and experience of designers using an interactive procedure. Sensory analysis (or sensory evaluation) is a scientific discipline that makes use of human senses (sight, smell, taste, touch, and hearing) for the purpose of evaluating consumer products based on certain evaluation criteria. Using a sensory evaluation method, 3D virtual try-on result can be accessed by designers through touch and observation based on their experience and professional knowledge. By utilizing a predefined pattern modification model, pattern can be modified automatically based on the evaluation results.

2 Experimental Work

In this research, a 3D-to-2D design-based personalized garment design process is proposed. Figure 1 presents the proposed design process. The proposed design process begins with a personalized garment block design from 3D-to-2D design method. Using the flattened garment block pattern, extension of the desired garment pattern can be realized considering the garment style and fabric property. Then, using a 3D virtual try-on procedure, a fast validation of the proposed pattern can be realized

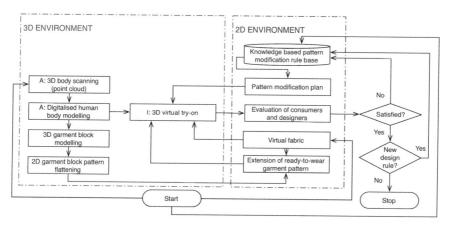

Fig. 1 Research framework and working process of the proposed design process

using virtual fabrics. Designers and consumers can be invited to participate in the evaluation section. His/her sensory evaluation result based on the predefined evaluation criteria will be input into a predefined knowledge-based pattern modification rule base. Pattern modification plans can be generated automatically. Using this method, 2D technical space of pattern making can be easily controlled by 3D perception space using 3D virtual try-on. Also, new design rule will be generated as a new input into the proposed knowledge-based pattern modification rule base in order to enhance the precision of the automatic pattern modification. As a knowledge-based learning system, the more designers and consumers involved in the decision making of the design process, higher satisfaction of the design solution can be ensured.

In order to realize the proposed design process, two experiments are proposed. In Experiment I, garment patterns for ready-to-wear garment products are extended from a set of garment block pattern designed by a 3D-to-2D design method. Fabric properties are tested in order to simulate corresponding virtual fabric. Based on the extended patterns and virtual fabric, a 3D virtual try-on is performed for further evaluation. In Experiment II, an automatic pattern modification rule base is defined. First, a set of evaluation criteria is defined, and then, corresponding pattern modification rules are discussed by pattern designers using their professional design knowledge.

2.1 Experiment I Production Pattern Design and 3D Virtual Try-on

In this study, a personalized garment block for a physically disabled lady with scoliosis from our previous study is applied. Using the proposed block patterns, the design case of a shirt can be obtained. The specifications design sheet in Fig. 2 describes

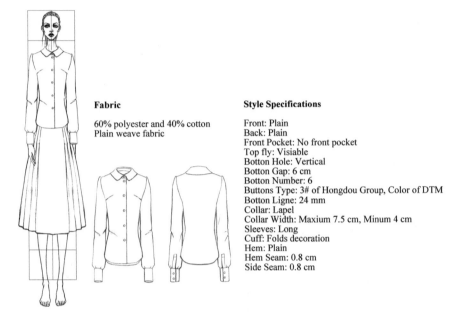

Fabric

60% polyester and 40% cotton
Plain weave fabric

Style Specifications

Front: Plain
Back: Plain
Front Pocket: No front pocket
Top fly: Visiable
Botton Hole: Vertical
Botton Gap: 6 cm
Botton Number: 6
Buttons Type: 3# of Hongdou Group, Color of DTM
Botton Ligne: 24 mm
Collar: Lapel
Collar Width: Maxium 7.5 cm, Minum 4 cm
Sleeves: Long
Cuff: Folds decoration
Hem: Plain
Hem Seam: 0.8 cm
Side Seam: 0.8 cm

Fig. 2 Design specifications sheet of the shirt

the requirements for the design elements of the shirt designed by designers in this study.

Then, several extensions and sizing procedure are performed in order to obtain the desired garment style in Fig. 2. Professional knowledge of pattern designers is utilized to support this process. These operations are:

Step 1: Adding 1 cm to the side seam, center front and back lines for all patterns, creating the buttonholes.

Step 2: Drawing vertical lines from the lowest armhole points to create the new side seam lines, increasing the length of the new side seam lines by 24.5 cm from the waistlines and drawing new lines vertical to the side seam lines to be the new bottom lines.

Step 3: Symmetrizing the back waist dart with the same value in the opposite direction of the original dart, while the symmetrical line is the end of dart legs (The two lines that converge at a predetermined point on the pattern.).

Step 4: Drawing a straight line to be the grainline (the center of the sleeve from top of curved top of the sleeve top to wrist level) of the sleeve, measuring the front and back armholes of the previous patterns, recording the measurements on the patterns for future reference, adding the front and back armhole measurements together, and dividing the value into four to be the cap height of the sleeve (the distance from widest line of the sleeve to the top at the grainline).

(a) (b) (c)

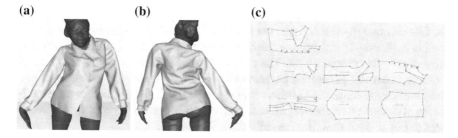

Fig. 3 Design result in both front and back views and the corresponding garment patterns

Step 5: Determining the length of the cuff as 20 cm (add 4 cm to the girth of artifice as 16 cm), the width of cuff as 5 cm.

Step 6: Determining the height of the stand collar as 3 cm and the height of the top collar as 4 cm.

By adding seam allowances and using *Modaris* software also with the virtual fabric, the production patterns of the designed shirt are finished as shown in Fig. 3.

2.2 Experiment II Evaluation and Adjustment of the 3D Try-on Perception

In this study, a session of sensory evaluation is realized by a group of fashion designers and pattern makers in order to quantitatively characterize the 3D virtual try-on perception of the shirt designed using the customized block patterns (Hong et al. 2018). Then, the adjustment of the shirt patterns can be realized in the 2D technical space according to the sensory evaluation results on the performance of the finished shirt in the 3D virtual try-on space. Evidently, the key issue of this adjustment is to set up a model characterizing the relationship between the technical space and 3D virtual product perceptual space. This model will permit to generate the appropriate technical parameters of the garment according to the desired values of sensory evaluation on the effects of 3D virtual garment try on (See Fig. 4).

In this study, the aim of identifying the shirt fit in 3D virtual try-on evaluation is to generate normalized sensory descriptors, which constitute the common communication language between fashion designers, pattern makers, and garment consumers.

Fig. 4 Model characterizing the relation between technical parameters and perception on 3D virtual garment try-on

Five experienced fashion designers are involved in the evaluation. The parameters of the fabric chosen by the designers as described in the design specifications sheet as well as the finished production patterns constitute the inputs to the *Modaris* software for realizing 3D virtual try-on. The style and design elements in the design specifications sheet can generate a common idea of the designers for the evaluation. In this context, the sensory evaluation results on the fit of the designed shirt, given by different designers, can be very close to each other. The sensory evaluation procedure used in our study is described as follows.

Each trained designer generates an exhaustive list of categories describing the shirt fit performance according to his/her professional knowledge. The three most relevant categories describing the key positions of the shirt have been selected: "*overall image*," "*fit in width*," and "*fit in details*." Then, a list of descriptors describing the shirt fit in different categories is generated by the designers using their garment design knowledge and pattern design knowledge.

After that, there is a section reducing redundant descriptors and those irrelevant to the fit of a shirt, by performing a "round table" discussion inside the panelists. This step leads to the generation of eight normalized descriptors describing the apparel fit performance (see Table 1). For each descriptor, a scale of five evaluation scores, ranging from −2 to 2, is also obtained. "−" means that the garment is tight or small related to the body shape while "+" means in the opposite direction (big or loose). 0 is a perfect fit on the wearer. Each score of the scale is defined semantically in Table 2.

Table 1 Sensory descriptors describing the apparel fit performance

Categories	Apparel fit performance descriptor
D_A: *Overall image*	D_{A1} *Overall fit*
	D_{A2} *Length*
D_B: *Fit in width*	D_{B1} *Waist fit*
	D_{B2} *Breast fit*
	D_{B3} *Hem fit*
D_C: *Fit in details*	D_{C1} *Shoulder fit*
	D_{C2} *Neck fit*
	D_{C3} *Arm hole fit*

Table 2 Evaluation scores, corresponding linguistic terms and TFNs

Evaluation scores	Linguistic terms
2	*Very loose/big (VL)*
1	*A little loose/big (AL)*
0	*Perfect (P)*
−1	*A little tight/small (AT)*
−2	*Very tight/small (VT)*

By repeating the evaluation two times and taking the average of the evaluation scores for each sensory descriptor, we finally obtain a matrix composed of all evaluation scores.

The adjustment of the current patterns will be realized using a rule-based model characterizing the relation between evaluations values on the shirt fit (perceptual space) and modifications of garment patterns (technical space). It has been established by exploiting the common professional knowledge of pattern makers through a round table discussion between these panelists. Five experienced pattern makers are involved in the production process. There are two steps for modeling the relationship:

Step 1: Identification of the shirt modification rules

These rules will enable to determine the key points or key lengths of the shirt production patterns corresponding to each sensory descriptor in order to make the final shirt very close to the target wearing effect wished by the designers. The final modification rules provided by the pattern makers are given in Table 3. Normally, for each sensory descriptor, there are several alternative modification rules. However, in practice, only one rule is applied during the adjustment. One example is given below.

If we wish to modify "overall fit" (D_{A1}), then we can change the length of either waistline (D_{A1a}), or breast line (D_{A1b}) or shoulder line (D_{A1c}).

Step 2: Identification of the new values of change for garment patterns

Table 3 Modification rules based on evaluation result

Sensory descriptors on garment fit	Rule code	Modification rules
D_{A1} Overall fit	D_{A1a}	Change the length of waistline
	D_{A1b}	Change the length of breast line
	D_{A1c}	Change the length of shoulder line
D_{A2} Length	D_{A2a}	Change the length of the garment
D_{B1} Waist fit	D_{B1a}	Change the cut of side seam
	D_{B1b}	Change the value of waist dart
D_{B2} Breast fit	D_{B2a}	Change the cut of side seam
	D_{B2b}	Change the value of breast dart
D_{B3} Hem fit	D_{B3a}	Change the cut of side seam
D_{C1} Shoulder	D_{C1a}	Change the slope of shoulder line
	D_{C1b}	Change the length of shoulder line
	D_{C1c}	Change the width of neckline
D_{C2} Neck	D_{C2a}	Change the width of neckline
	D_{C2b}	Change the depth of neckline
D_{C3} Armhole	D_{C3a}	Change the position of sleeve top
	D_{C3b}	Change the curvature of the armhole

Table 4 Values of change for shirt production patterns

Evaluation scores	−2 (cm)	−1 (cm)	0	1 (cm)	2 (cm)
D_{A1a}	+8	+4	0	−4	−8
D_{A1b}	+8	+4	0	−4	−8
D_{A1c}	+4	+2	0	−2	−4
D_{A2a}	+4	+2	0	−2	−4
D_{B1a}	+3	+1	0	−1	−3
D_{B1b}	+2	+1	0	−1	−2
D_{B2a}	+3	+1	0	−1	−3
D_{B2b}	+2	+1	0	−1	−2
D_{B3a}	+3	+1	0	−1	−3
D_{C1a}	+4°	+2°	0	−2°	−4°
D_{C1b}	+3	+1	0	−1	−3
D_{C1c}	+3	+1	0	−1	−3
D_{C2a}	+3	+1	0	−1	−3
D_{C2b}	+3	+1	0	−1	−3
D_{C3a}	+2	+1	0	−1	−2
D_{C3b}	+2	+1	0	−1	−2

For each modification rule, the change of the identified key point or key length is determined according to the evaluation score of the corresponding sensory descriptor. The whole values of change of patterns related to all the modification rules, provided by the designers, are shown in Table 4. For example, when applying the modification rule D_{A1a} related to the "overall fit," if the evaluation score is −1 (a little tight), then 4 cm will be added to the width of waistline in horizontal direction.

In practice, one modification rule and its corresponding pattern changing value can be arbitrarily selected and its try-on result can be quantitatively characterized using the sensory evaluation. If the adjustment result is not satisfying, another rule of the same sensory descriptor will be selected in order to generate a new try-on result. This procedure can be carried out repeatedly until finding the most relevant adjustment plan.

By using the previous two steps, we set up the relationship between 3D virtual shirt try-on results and 2D pattern parameters (key points and key lengths). This rule-based model permits to reach a desired perception of shirt fit by adjusting the 2D pattern parameters. In this study, the procedure of *Design–Display–Evaluation–Adjustment* with the model can be performed repeatedly until a satisfying design solution is obtained, as shown in Fig. 5 (Chen et al. 2015). D_{A2} *Length*, D_{B1} *Waist fit*, D_{B2} *Breast fit*, D_{B3} *Hem fit*, D_{C1} *Shoulder fit*, and D_{C2} *Neck fit* of the initial pattern are modified using the rule-based model based on the perception of the involved designers.

To validate the design result, following the design specifications sheet, the garment is produced following the garment patterns after modification. To evaluate the real

(a) **(b)** **(c)**

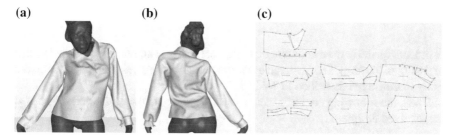

Fig. 5 Design result after modification in both front and back views and the corresponding garment patterns

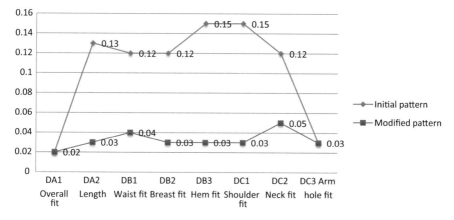

Fig. 6 Distances of aggregated evaluation results of all evaluation criteria to the degree of "Perfect (1.5, 2, 2.5)"

shirt produced using the pattern modified as the evaluation results before, a fitting procedure is proposed. The fitting result is shown in Fig. 6. As for the customer, the look and comfort of the shirt are perceived as fine during the fitting.

In order to validate the proposed design process, another group of designers is invited to compare the initial garment virtual try-on result and the final modified garment virtual try-on result using the evaluation criteria presented in Table 1.

First, there is a training section about the purposed of the evaluation. Then both the initial garment virtual try-on result and the final modified garment virtual try-on are presented to the invited designer group. The invited designers are free to operate the computer to observe the virtual try-on results. Each of the invited designers is assigned to finish the evaluation independently without any discussion with other designers.

3   Results and Discussion

In order to quantify the evaluation degrees, a set of fuzzy numbers is assigned to each of the linguistic term. The involved evaluation linguistic term and their corresponding fuzzy numbers are described in Table 5.

Then, the following equation can be used to aggregate the evaluation result of all the involved designers:

$$a_{ij} = \left(\frac{1}{m} \sum_{j=1}^{l} a_{ijh} t_1, \frac{1}{m} \sum_{j=1}^{l} a_{ijh} t_2, \frac{1}{m} \sum_{j=1}^{l} a_{ijh} t_3 \right)$$

where t_1, t_2, t_3 correspond to the value of TFNs, and their values are taken according to Table 3 (Hong et al. 2018).

Then, Euclidean distance of between all the aggregated evaluation results to the semantic degree "Perfect (1.5, 2, 2.5)" is calculated in order to measure the satisfaction of invited designers of both initial garment virtual try-on result and finial modified virtual try-on result, as presented in Fig. 6. These distances indicate the satisfaction degree of each part of the virtual try-on result in terms of different evaluation criteria. Shorter distance indicates higher membership degree.

D_{A2} Length, D_{B1} Waist fit, D_{B2} Breast fit, D_{B3} Hem fit, D_{C1} Shoulder fit, and D_{C2} Neck fit are less satisfactory compared to the initial pattern. Some modification should be performed on these parts, matching the designers' idea referring to Experiment II.

Generally, the modified pattern is more "perfect" compared with the initial pattern, which indicates that the proposed *Design–Display–Evaluation–Adjustments* procedure is able to help to reach a desired perception of shirt fit by adjusting the 2D pattern parameters using the proposed rule-based model. 2D pattern parameters (key points and key lengths) can be adjusted by the evaluation result of the 3D virtual try-on results.

Table 5 Linguistic rating scale and corresponding fuzzy numbers

Linguistic values	TFNs
Very loose/big	(2.5, 3, 3.5)
A little loose/big	(2, 2.5, 3)
Perfect	(1.5, 2, 2.5)
A little tight/small	(1, 1.5, 2)
Very tight/small	(0.5, 1, 1.5)

4 Conclusions

This chapter demonstrates the application of artificial intelligence for fashion design evaluation and automatic pattern modification. In this paper, to validate the applicability of 3D-to-2D garment design method, we introduced an application case of designing a personalized garment for physically disabled people with scoliosis (PDPS) using a set of garment block pattern designed by a 3D-to-2D garment design method. The proposed design process begins with a personalized garment block design from 3D-to-2D design method. Desired garment patterns are realized by the extension of the previous garment block pattern considering both desired garment style and fabric properties. After that, using the proposed pattern and virtual fabric, a 3D virtual try-on can be realized and ensures a fast validation of the proposed pattern. Designers and consumers are able to participate in the evaluation section. Also, we defined a rule-based pattern modification model to validate the designer design idea using artificial intelligence tools. Based on sensory evaluation results, pattern modification plan can be generated. Though a validation experiment, it can be proved that, as knowledge-based learning process, the proposed design process is able to ensure a high level of design satisfaction of consumers.

Bibliography

Chen X, Tao X, Zeng X, Koehl L, Boulenguez-Phippen J (2015) Control and optimization of human perception on virtual garment products by learning from experimental data. Knowl-Based Syst 87:92–101

Hong Y, Bruniaux P, Zeng X, Curteza A, Liu K Design and evaluation of personalized garment block design method for atypical morphology using the knowledge-supported virtual simulation method. Text Res J 0040517517708537. https://doi.org/10.1177/0040517517708537

Hong Y, Bruniaux P, Zeng X, Liu K, Curteza A, Chen Y Visual-simulation-based personalized garment block design method for physically disabled people with scoliosis (PDPS). Autex Res J. https://doi.org/10.1515/aut-2017-0001

Hong Y, Curteza A, Zeng X, Bruniaux P, Chen Y (2016a) Sensory evaluation based fuzzy AHP approach for material selection in customized garment design and development process. In: Book of abstracts, vol 133. IOP Publishing, Iasi, pp 1–8

Hong Y, Yang M, Chen Y (2013) Development of clothing micro climate monitoring system for human physiological indexes. J Text Res 1:020

Hong Y, Zeng X, Bruniaux P (2016b) Knowledge acquisition and modeling of garment product development In: Uncertainty modelling in knowledge engineering and decision making: proceedings of the 12th international FLINS conference (FLINS 2016), vol 10. World Scientific, Roubaix, pp 438–444

Hong Y, Zeng X, Bruniaux P (2016c) Selection and application of key performance indicators for design and production process. In: Uncertainty modelling in knowledge engineering and decision making: proceedings of the 12th international FLINS conference. World Scientific, Roubaix, pp 1008–1014

Hong Y, Zeng X, Bruniaux P, Curteza A, Chen Y (2017a) Movement analysis and ergonomic garment opening design of garment block patterns for physically disabled people with scoliosis using fuzzy logic. In: International conference on applied human factors and ergonomics. Springer, pp 303–314

Hong Y, Zeng X, Bruniaux P, Liu K (2017b) Interactive virtual try-on based three-dimensional garment block design for disabled people of scoliosis type. Text Res J 87(10):1261–1274. https://doi.org/10.1177/0040517516651105

Hong Yan XZ, Bruniaux P, Liu K, Chen Y, Zhang X (2017c) Collaborative 3D-to-2D tight-fitting garment pattern design process for scoliotic people. Fibres Text East Eur 25(5):113–118

Hong Y, Zeng X, Wang Y, Bruniaux P, Chen Y (2018) CBCRS: an open case-based color recommendation system. Knowl-Based Syst 141:113–128

Wang L, Zeng X, Koehl L, Chen Y (2015) Intelligent fashion recommender system: fuzzy logic in personalized garment design. IEEE Trans Hum-Mach Syst 45(1):95–109

Yan Hong PB, Zeng Xianyi, Liu Kaixuan, Chen Yan, Dong Min (2017) Virtual reality based collaborative design method for designing customized garment of disabled people with scoliosis. Int J Cloth Sci Technol 29(2):226–237. https://doi.org/10.1108/IJCST-07-2016-0077

Garment Wearing Comfort Analysis Using Data Mining Technology

Kaixuan Liu

Abstract Clothing pressure is an important factor that influences wear comfort. It can reflect wear comfort intuitively. With the development of virtual-reality technology, the numerical clothing pressures were widely applied in the evaluations of garment wear comfort. Compared with traditional measuring methods of garment pressure, the virtual clothing pressure-measuring method has the advantages of lower cost, time saving, efficiency improving, easy accessibility, and higher accuracy. In this article, sixty human's postures in daily life were designed using CLO 3D Modelist. Next, the numerical clothing pressure data were collected, respectively, under the sixty postures in a virtual environment. Finally, factor analysis was applied to process the data. The result shows that the pants' pressure wear comfort is mainly influenced by four factors; there are thigh–hip factor, shank factor, waist factor, and crotch factor. The share of these four factors' influence accounts for about 82.0%, and the thigh–hip factor takes up the biggest share and reaches to 44.5%.

Keywords Numerical clothing pressure · Wearing comfort · Virtual-reality
Factor analysis

1 Introduction

Garment wearing comfort is always consumers' concern. No matter how beautiful a garment is, and how excellent the fabric's properties are, the customers will not select the garment if the wearing comfort is bad. Many factors influence wearing comfort, such as ease allowance (Chen et al. 2006, 2008; Thomassey and Bruniaux

K. Liu (✉)
Apparel and Art Design College, Xi'an Polytechnic University, Xi'an 710048, China
e-mail: liukaixuan819@163.com

K. Liu
University of Lille 1, Nord de France, 59000 Lille, France

K. Liu
GEMTEX Laboratory, ENSAIT, 59100 Roubaix, France

© Springer Nature Singapore Pte Ltd. 2018
S. Thomassey and X. Zeng (eds.), *Artificial Intelligence for Fashion
Industry in the Big Data Era*, Springer Series in Fashion Business,
https://doi.org/10.1007/978-981-13-0080-6_13

2013), fabric properties (Barker 2002), garment pattern making (Jeong et al. 2006). Currently, research studies of wearing comfort mainly focus on two aspects: physiological wearing comfort and psychological wearing comfort. The research methods of wearing comfort are divided into quantitative analysis and qualitative assessment. As wearing comfort is a complex sense, Zeng et al. proposed to evaluate garment and fabric products using fuzzy logic, artificial intelligence, and sensory evaluation techniques (Zeng and Koehl 2003; Zeng et al. 2004, 2008). Hollies et al. (1979) proposed a human perception analysis approach to clothing comfort. However, the previous studies are mainly focused on qualitative analysis, and few studies involved quantitative research.

Many factors influence garment wearing comfort. Even, different human body parts influence wearing comfort differently in different conditions. For example, the crotch has more obvious impact on wearing comfort than the shank. Many people complain that the waists and crotches are not comfort when they wear unfit pants. On the contrary, few customers complain uncomfortable on the parts below knee. Often, most people can easily tell in which parts they feel uncomfortable or comfort when the garments are being tried on. However, why they feel uncomfortable or comfortable in different parts? Few research studies involved in these aspects. In this research, we try to use mathematics analysis methods to explain these questions.

Clothing pressure is one of the most important elements influencing wearing comfort (Kilinc-Balci 2011; Liu and Chen 2015; Bragança et al. 2015; Liu et al. 2013). As clothing pressures reflect wearing comfort intuitively, whether to change the fabric mechanical properties, garment ease allowance, fashion style, or garment pattern, the final changes are clothing pressures. Therefore, clothing pressure is a key indicator reflects wearing comfort (Morooka et al. 2005; Dongsheng and Qing 2003; You et al. 2002a, b; Tsujisaka et al. 2004; Kamalha et al. 2013). In this context, we selected clothing pressure as the key analysis index to analyze garment wearing comfort.

The traditional measuring method of clothing pressure requires an expensive pressure-measuring device. All pressure sensors of this device are linked with the host by wires. These pressure sensors are easy to mismatch or fall out during strenuous activity. Moreover, the wires limit free activities. With the development of three-dimensional (3D) virtual-reality technology, the numerical clothing pressures were widely applied in evaluation of garment wearing comfort (Dan et al. 2016; Zhang et al. 2015; Wong et al. 2004). Zhang et al. (2015) pointed out that numerical simulations of the clothing pressure distribution are critical to the optimal design of apparel products. The research of Wong et al. (2004) showed that the numerical simulation method of clothing pressures can provide reliable prediction in terms of pressure comfort. Compared with traditional measuring methods of garment pressure, the measuring method of numerical clothing pressure has the advantages of lower cost, time saving, efficiency improving, easy accessibility, and higher accuracy (Fan and Chan 2005). Finally, the numerical clothing pressure was selected to analyze garment wearing comfort.

In this research, we used the 3D virtual-reality technology to simulate human daily activities. Then, we proposed a new method to measure dynamic clothing pressure.

The measuring points of clothing pressure were arranged in garment pattern. Next, the virtual clothing pressures were measured according to these arranged points. Finally, the data mining technology was applied to process the collected data. The first aim of this research was to find which parts of human body were influenced by clothing pressure significant. The second aim was to provide a theoretical foundation for the evaluation of garment wearing comfort. In order to realize the proposed method, we organized the structure of this article as follow: In Sect. 2, we introduced a novel measuring method to collection numerical clothing pressures. In Sect. 3, test of outliers, descriptive analysis, and factor analysis were applied to the collected data. Some conclusions and possible further works were presented in Sect. 4.

2 Method

Currently, the development of garment productions focuses on static condition, but rarely involves dynamic condition. That is why many customers complain that they always cannot find form-fitting pants. Thus, we designed an experiment to evaluate the garment dynamic wearing comfort and tried to find which factors influence the garment dynamic wearing comfort. The evaluation object of this research was a pair of pants, which fit a wearer perfectly in static and dynamic conditions. A 3D digital human model replaced the wearer to carry out the experiment. Therefore, we set the model's body dimensions equal to the wearer's body dimensions. Note that there are some differences between real and virtual clothing pressures. As the clothing pressures of this research were measured in a virtual environment, the unit of clothing pressure kPa refers to virtual kPa in the following sections.

2.1 Action Design for Measuring Clothing Pressures

With the changes of actions, clothing pressures fluctuate continually. This phenomenon indicates people's activities influence cloth pressures significantly. Wearing comfort is influenced by not only static clothing pressure, but also dynamics clothing pressure. Thus, the analysis of wearing comfort involved both static and dynamic clothing pressures. The actions, which people do frequently in daily life and work, such as, walk, sit, ran, and squat, were design for an avatar by *CLO 3D Modelist software* (Fig. 1). Hundred and twenty postures, which were extracted from these actions, were used for measuring clothing pressures.

Fig. 1 Action design of daily life and work for measuring clothing pressures

2.2 Measurement of Clothing Pressures

In this section, we proposed a new method to measure numerical clothing pressures. As fabric properties influence clothing pressure significantly (You et al. 2002a; Wong et al. 2004; Zhang et al. 2002), in order to avoid the influence of different fabric properties on garment wearing comfort, we only selected a fabric in this study, whose mechanical properties are shown in Table 1. Generally, three steps are required to measure the clothing pressure by *CLO 3D Modelist* software.

Firstly, 33 measuring points were mapped on a pair of pants' patterns. As shown in Fig. 2a, measuring points *W1*, *W2*, … *W5* were equally distributed in the garment pattern of waistband piece; measuring points *F1*, *F2*, … *F18* were equally distributed in the garment pattern of front piece; *B1*, *B2*, … *B10* were equally distributed in the garment pattern of back piece. Note: As the pants' patterns are bilateral symmetry, we only arranged measuring points on half of the patterns.

Secondly, the patterns were tried on an avatar in the virtual environment (Fig. 2b). Then, we let the avatar do various motions, which were designed in Fig. 1; in the meanwhile, we, respectively, measured dynamic clothing pressures every two seconds according to the predefined measuring points (Fig. 2c).

Finally, 120 sets of clothing pressure data were collected, and each set of clothing pressure data contains 33 values (33 measuring points' clothing pressures). The whole data were applied to the following data mining.

Table 1 Fabric mechanical properties for virtual try-on and garment pressure measurement

Buckling stiffness-weft	Buckling stiffness-warp	Buckling ratio-weft	Internal damping	Friction coefficient	Buckling ratio-warp
30	30	50	1	3	50
Bending-weft	Bending-warp	Shear	Stretch-weft	Stretch-warp	Density
35	35	23	32	32	35

(a)   **(b)**   **(c)**

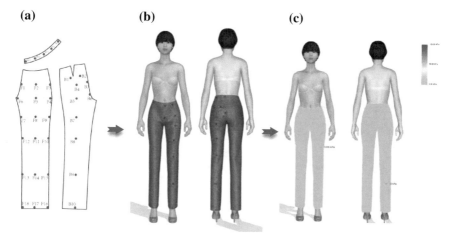

Fig. 2 Numerical clothing measuring method

3 Results and Discussion

3.1 Data Preprocessing and Analysis

As data are generally incomplete in the real world; it is inapposite to analyze data directly without preprocessing (Han et al. 2011). There are many methods to preprocess raw data. In this research, test of outliers and descriptive analysis were applied to preprocess the collected data.

3.1.1 Test of Outliers

Through the method of 3σ-rule (Han et al. 2011), we detected that some outliers existed in seven sets of clothing pressures. The presence of outliers can lead to an anamorphic analysis results; therefore, we excluded these seven sets of clothing pressure data. Finally, 113 samples were available for the subsequent analysis.

3.1.2 Descriptive Analysis

Descriptive analysis is a practical approach that is simple, direct, and effective to reflect some data features. The results of descriptive analysis are shown in Table 2, the range of clothing pressure at measuring points *F6, F10, F12,* and *B8* is quite large, especially, and the range of point F6 reaches to 80.89 kPa. It indicates that the wearing comfort on the parts of knee and crotch bottom was influenced significantly during activity (The positions of points F6, F10, F12, and B8 correspond with human

Table 2 Descriptive statistics of anthropometric measurements (unit: kPa)

No.	Point	Sample number	Minimum	Maximum	Mean	Standard deviation	Coefficient variation (%)
1	F1	112	2.28	19.54	7.57	3.26	43.14
2	F2	112	0.86	23.15	9.86	5.44	55.13
3	F3	112	2.82	27.32	8.21	4.41	53.69
4	F4	112	2.00	17.00	7.16	2.84	39.64
5	F5	112	1.03	20.48	7.55	4.07	53.95
6	F6	112	15.10	95.99	35.43	12.62	35.62
7	F7	112	2.70	31.83	9.07	4.75	52.36
8	B1	112	4.41	17.35	8.84	2.46	27.79
:	:	:	:	:	:	:	:
33	B10	112	0.01	9.63	1.12	1.79	160.08

body knee and crotch bottom, see Fig. 2.) The maximums of clothing pressures at points $W1$, $W2$, $F6$, $F10$, $F11$, $F12$, $B6$, and $B8$ are over 40 kPa; especially, point F6 reaches upward of 90 kPa. It indicates that the garment's wearing comfort at waist, knee, and crotch bottom are worse than other parts (The positions of points $W1$, $W2$, $F6$, $F10$, $F11$, $F12$, $B6$, and $B8$ correspond with human body waist, knee, and crotch bottom, see Fig. 2). Designers should modify these parts' patterns if the wearer needs better wearing comfort. Coefficient variations of points $F10$, $F11$, $F12$, $F14$, $F16$, $F15$, $F17$, $B8$, $B9$, and $B10$ exceed 80%; especially, points F17 and B10 are over 160%. It indicates the degree of the variation of clothing pressure at knee and below knee is significant (The positions of points $F10$, $F11$, $F12$, $F14$, $F16$, $F15$, $F17$, $B8$, $B9$, and $B10$ correspond with human knee and below knee see Fig. 2). However, the mean values of clothing pressures below knee are small (Table 2); this part has no significant influence on wearing comfort.

3.2   Factor Analysis

Factor analysis is a data mining method used to describe variability among observed, correlated variables in terms of a potentially lower number of unobserved variables called factors. From the previous discussion, we know that every human body part influences garment wearing comfort differently. However, some parts influence wearing comfort significantly, and some parts have little impact on wearing comfort. Therefore, it is inappropriate to consider every human body part has the same effect weight on wearing comfort. In this research, we arranged 33 measuring points on pants' patterns. Nevertheless, it is too many measuring points to analyze pressure wearing comfort directly. Therefore, we explored factor analysis, which is widely

Table 3 KMO and Bartlett's test

Kaiser–Meyer–Olkin	Kaiser–Meyer–Olkin measure of sampling adequacy	0.813
Bartlett's test of sphericity	Significance	0.000

applied to data reduction or structure detection (Comrey and Lee 2013), to process the collected data. The aim of factor analysis was to detect how different human body parts influence wearing comfort.

3.2.1 Kaiser–Meyer–Olkin Test and Bartlett's Test of Sphericity

Kaiser–Meyer–Olkin (KMO) (Cerny and Kaiser 1977; Dziuban and Shirkey 1974; Kaiser 1970) and Bartlett's test (Dziuban and Shirkey 1974) are applied to test whether factor analysis is appropriate for data processing in this research. In Table 3, the KMO value is about 0.813, indicating that factor analysis can be meritoriously conducted (Kaiser and Rice 1974); the significance (sig) value is 0, indicating that the hypothesis of independence is rejected, thereby providing evidence that factor analysis may be suited to the data (Dziuban and Shirkey 1974).

3.2.2 Principal Factors Extraction and Explanation

Scree plot whose X-axis represents the series number of component and whose Y-axis represents the eigenvalue described the result of factor analysis. As shown in Fig. 3, eigenvalues of the first seven components exceed one and other components' eigenvalues approach zero gradually. Thus, we tried to extract the first seven components for subsequent analysis (Table 4). If their accumulative contribution was low, we continued to add components until the accumulative contribution was enough to reflect the overall information of clothing pressure. Actually, we can extract 33 components at most due to the existence of 33 measure points and rank them according to their contribution rates. However, most of the low-ranked components are difficult to explain, and there is no need to use them in real applications because their contribution rates are very slight. Only the highest ranked components are interesting for us.

According to the rotated factors loading of the extracted seven components in Table 4, we defined five factors, which influence wearing comfort significantly, as follows:

The first and the fifth components mostly reflect the effect of human body waist and hip on wearing comfort, mainly including *W1*, *W2*, *W3*, *W4*, *W5*, *F2*, *F3*, *F4*, *B1*, *B2*, *B3*, *B4,* and *B5*. The positions of these points correspond with human body front waist and hip (see Fig. 2); therefore, we denominate it "waist–hip factor."

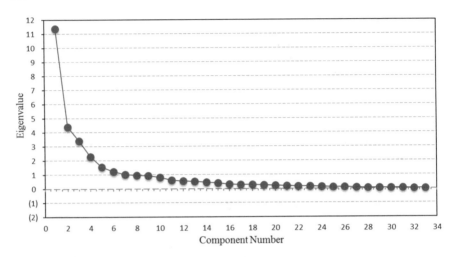

Fig. 3 Scree plot on component analysis

Table 4 Rotated component matrix of clothing pressures

No.	Point	Component						
		1	2	3	4	5	6	7
6	F1	0.33	0.15	0.59	0.00	0.23	−0.16	−0.01
7	F2	0.65	0.13	0.49	−0.18	−0.05	0.09	0.11
8	F3	0.68	0.24	0.33	0.01	−0.14	0.13	−0.05
9	F4	0.66	0.20	0.40	0.06	−0.17	0.07	0.16
10	F5	0.41	−0.01	0.68	−0.10	0.01	0.19	0.21
⋮	⋮	⋮	⋮	⋮	⋮	⋮	⋮	⋮
33	B10	0.11	0.53	0.07	0.51	0.12	−0.14	0.50

The second and seventh components mostly reflect the effect of human body knee and shank on wearing comfort, mainly including $F10$, $F11$, $F12$, $F13$, $F14$, $F15$, $F16$, and $F18$. The positions of these points correspond with human body knee and shank (see Fig. 2); therefore, we denominate it "knee–shank."

The third and sixth components mostly reflect the effect of human crotch on wearing comfort, mainly including $F5$ and $B6$. The position of these points corresponds with human body crotch (see Fig. 2); therefore, we denominate it "crotch factor."

The fourth component mostly reflects the effect of human body back thigh and calf on wearing comfort, mainly including $B7$ and $B9$. The positions of the two points correspond with human body back thigh and back calf (see Fig. 2); therefore, we denominate its "back thigh–calf factor."

The total contribution rate of accumulated variance of the first four factors (the first seven components) reaches 76.48% (Table 5). It indicates that the waist–hip factor, knee–shank factor, crotch factor, and back thigh–calf factor can represent the

Table 5 Total variance explained

Factor	% of variance	Cumulative %	Total	Component
Waist–hip factor	39.17	39.17	12.93	1
Knee–shank factor	16.4	55.57	5.41	2
Crotch factor	13.96	69.53	4.61	3
Thigh–calf factor	6.95	76.48	2.3	4

information of whole data in general. From the contribution rates of accumulated variance of the first five factors, we know that the most significant impact on wearing comfort is the waist–hip factor. Its contribution rate of accumulated variance reaches to 39.17% (Table 5). Furthermore, the knee–shank factor and crotch factor also account for bigger proportions (together accounting for 30.36%, see Table 5). Summarized the above analyses, we derived that the wearing comfort of pants is mainly influenced by the parts of hip, waist, knee, shank, and crotch, especially waist and hip.

The weight values are usually assigned by Delphi method (expert scoring methods) in the field of Kansei Engineering (Huang et al. 2011). However, this method is influenced significantly by subjective factor. In this research, the weight values of waist–hip factor, knee–shank factor, crotch factor, and back thigh–calf factor, which influence wearing comfort, are 39.17, 16.40, 13.96, and 6.95% (Table 5); the result provides an important reference for researchers to assign the weight value in wearing comfort evaluation.

3.3 Wearing Comfort Analysis on Different Human Body Parts

3.3.1 The Wearing Comfort Analysis on Waist and Hip

During activity, the shape of hip and waist changes obviously. Factor analysis indicated the contribution rate of variance of hip–waist factor accounts for 39.17%; therefore, hip and waist are the emphases to analyze the wearing comfort. Figure 4 shows that clothing pressures mainly concentrate upon below 20 kPa. Overall, the clothing pressures at waist and hip fluctuate fiercely, especially front waist (points W1, W2, and W3) and back central hip (points B4 and B5) (Fig. 4a, c). It indicates that the wearing comfort deteriorates in dynamic conditions. Thus, fashion designers should focus on waist and hip, especially front waist and back central hip if they want to improve pants' wearing comfort. Moreover, according to the fluctuation range of clothing pressure in Fig. 4, we suggested that the wearing comfort is acceptable if numerical clothing pressures are below 20 kPa at the parts of hip and waist. If the clothing pressures of these parts exceed 20 kPa, the wearing comfort may become worse; the bigger the clothing pressures are, the worse the wearing comfort will be.

3.3.2   The Wearing Comfort Analysis on Knee and Shank

The factor analysis indicated the contribution rate of variance of knee and shank factor accounting for 16.40%. As shown in Fig. 5, the fluctuation range of clothing pressures at points *F13*, *F14*, *F15*, *F16,* and *F18* is mainly below 10 kPa. On the contrary, at point *F10*, *F11* and *F12* are above 10 kPa, even more than 70 kPa. The result indicates that the wearing comfort becomes worse at knee (point *F10*, *F11,* and *F12*) and has little change at shank (points *F13*, *F14*, *F15*, *F16,* and *F18*) in dynamic condition. Thus, fashion designers should focus on not shank but knee if they want to improve pants' wearing comfort. In addition, according to the fluctuation range of clothing pressures in Fig. 5, we suggested that the wearing comfort is acceptable if numerical clothing pressures are below 10 kPa at the parts of knee and shank. If the clothing pressures of these parts exceed 10 kPa, the wearing comfort may become worse; the bigger the clothing pressures are, the worse the wearing comfort will be.

3.3.3   The Wearing Comfort Analysis on Crotch

The factor analysis indicated the contribution rate of variance of crotch factor accounting for 13.96%. As shown in Fig. 6, the clothing pressures at point *F5* mainly fluctuate below 20 kPa. On the contrary, the clothing pressures at point *F6* mainly fluctuate above 20 kPa, even as high as about 100 kPa. The result indicates that the wearing comfort becomes worse at crotch bottom (point *F5*) in dynamic condition. This part is one of the most important zones for garment wearing comfort. Thus, fashion designers should pay more attention to crotch bottom, especially sportswear design. In addition, according to the fluctuation range of clothing pressure in Fig. 6, we suggest that the wearing comfort is acceptable if numerical clothing pressures are below 20 kPa at the part of crotch. If the clothing pressures of this part exceed 20 kPa, the wearing comfort may become worse; the bigger the clothing pressures are, the worse the wearing comfort will be.

3.3.4   The Wearing Comfort Analysis on Back Thigh and Back Calf

The factor analysis indicated the contribution rate of variance of back thigh–calf factor accounting for 6.95%. As shown in Fig. 7, the clothing pressures at back thigh and back calf mainly concentrate below 10 kPa, and the biggest value of clothing pressure is only about 20 kPa. The result indicates that parts of back thigh and back calf have no significant effect on wearing comfort. In addition, according to the fluctuation range of clothing pressure in Fig. 7, we suggest that the wearing comfort is acceptable if numerical clothing pressures are below 10 kPa at the part of back

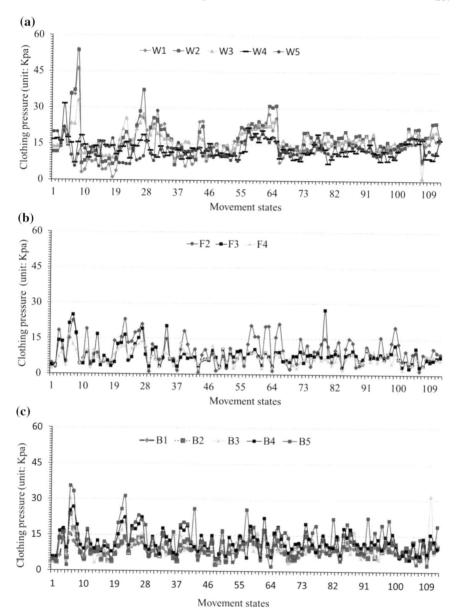

Fig. 4 Clothing pressures on back hip at different movement states

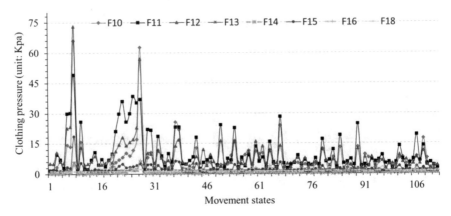

Fig. 5 Clothing pressures on knee and shank at different movement states

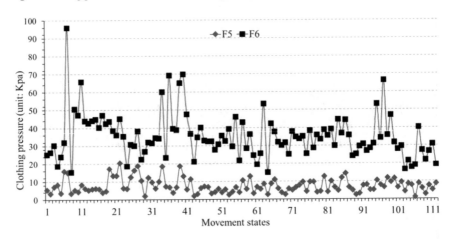

Fig. 6 Clothing pressures on crotch at different movement states

thigh and back calf. If the clothing pressures of these two parts exceed 10 kPa, the wearing comfort may become worse; the bigger the clothing pressures are, the worse the wearing comfort will be.

3.4 Limitation

Though the present study tries to do comprehensive research on wearing comfort from the aspect of numerical clothing pressure, the conclusions, which are based on a limited activity condition in daily life and work, are only for evaluating and developing daily clothing. Due to the change in diversity of clothing pressure in different activity states, like cycling, skiing, gymnastics, field events, the in-depth

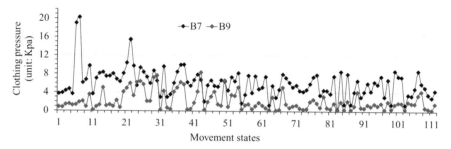

Fig. 7 Clothing pressures on side hip at different movement states

research on wearing comfort of professional sports remains a subject for further study. Moreover, the clothing pressures (numerical clothing pressure) were measured in a 3D virtual environment, and the scope of application of the results limits the same condition.

4 Conclusions and Prospects

In this research, we used 3D virtual-reality technology to simulate human activities and proposed a novel method to measure numerical clothing pressures. In addition, the data mining technology was applied to process the collected data. Compared to previous qualitative analysis of garment wearing comfort, we proposed a quantitative analysis of garment wearing comfort from the aspect of numerical clothing pressure. Summarized the analyses, we conducted three main conclusions:

- The wearing comfort of pants is mainly influenced by four factors: waist–hip factor, knee–shank factor, crotch factor, and back thigh–calf factor. Contribution rates of these four factors are 39.17%, 16.4%, 13.96%, and 6.95%, respectively.
- The parts of human body, which influence wearing comfort significantly, are hip, waist, crotch, and knee. The part below knee and the part of back thigh have no obvious effect on wearing comfort.
- The wearing comfort is acceptable if the numerical clothing pressures are below 20 kPa at the parts of hip, waist, and crotch and below 10 kPa at the parts of back thigh, knee, and shank.

All of the analysis about the wearing comfort of pants services optimal structure designing of pants' patterns and quantitative evaluation of garment wearing comfort. In addition, the proposed method also can be applied to other garment types, especially professional sportswear.

Acknowledgements This paper was financially supported by China National Endowment for the Arts.

References

Barker RL (2002) From fabric hand to thermal comfort: the evolving role of objective measurements in explaining human comfort response to textiles. Int J Cloth Sci Technol 14(3/4):181–200. https://doi.org/10.1108/09556220210437158

Bragança S, Fontes L, Arezes P, Edelman ER, Carvalho M (2015) The impact of work clothing design on workers' comfort. Procedia Manuf 3:5889–5896. https://doi.org/10.1016/j.promfg.2015.07.898

Cerny BA, Kaiser HF (1977) A study of a measure of sampling adequacy for factor-analytic correlation matrices. Multivar Behav Res 12(1):43–47. https://doi.org/10.1207/s15327906mbr1201_3

Chen Y, zeng X, Happiette M, Bruniaux P, Ng R, Yu W (2006) Estimation of ease allowance of a garment using fuzzy logic. In: Fuzzy applications in industrial engineering. Springer, pp 367–379. https://doi.org/10.1007/3-540-33517-x_15

Chen Y, Zeng X, Happiette M, Bruniaux P, Ng R, Yu W (2008) A new method of ease allowance generation for personalization of garment design. Int J Cloth Sci Technol 20(3):161–173. https://doi.org/10.1108/09556220810865210

Comrey AL, Lee HB (2013) A first course in factor analysis. Psychology Press, Hlilsdale, NJ

Dan R, X-r Fan, Shi Z, Zhang M (2016) Finite element simulation of pressure, displacement, and area shrinkage mass of lower leg with time for the top part of men's socks. J Text Inst 107(1):72–80. https://doi.org/10.1080/00405000.2015.1007621

Dongsheng C, Qing Z (2003) A study on clothing pressure for men's suit comfort evaluation. Int J Cloth Sci Technol 15(5):320–334. https://doi.org/10.1108/09556220310492598

Dziuban CD, Shirkey EC (1974) When is a correlation matrix appropriate for factor analysis? Some decision rules. Psychol Bull 81(6):358. https://doi.org/10.1037/h0036316

Fan J, Chan AP (2005) Prediction of girdle's pressure on human body from the pressure measurement on a dummy. Int J Cloth Sci Technol 17(1):6–12. https://doi.org/10.1108/09556220510577925

Han J, Kamber M, Pei J (2011) Data mining: concepts and techniques: concepts and techniques, 3rd edn. Morgan Kaufmann Publishers, Waltham, MA

Hollies NR, Custer AG, Morin CJ, Howard ME (1979) A human perception analysis approach to clothing comfort. Text Res J 49(10):557–564. https://doi.org/10.1177/004051757904901001

Huang M-S, Tsai H-C, Huang T-H (2011) Applying Kansei engineering to industrial machinery trade show booth design. Int J Ind Ergon 41(1):72–78. https://doi.org/10.1016/j.ergon.2010.10.002

Jeong Y, Hong K, Kim S-J (2006) 3D pattern construction and its application to tight-fitting garments for comfortable pressure sensation. Fiber Polym 7(2):195–202. https://doi.org/10.1007/BF02908267

Kaiser HF (1970) A second generation little jiffy. Psychometrika 35(4):401–415. https://doi.org/10.1007/BF02291817

Kaiser HF, Rice J (1974) Little Jiffy, Mark IV. Educ Psychol Meas 34(1):111–117. https://doi.org/10.1177/001316447403400115

Kamalha E, Zeng Y, Mwasiagi JI, Kyatuheire S (2013) The comfort dimension; a review of perception in clothing. J Sens Stud 28(6):423–444. https://doi.org/10.1111/joss.12070

Kilinc-Balci F (2011) How consumers perceive comfort in apparel. In: Song G (ed) Improving comfort in clothing, 1st edn. Woodhead Publishing, Cambridge, UK, pp 97–113

Liu Y, Chen D (2015) An analysis on EEG power spectrum under pressure of girdle. Int J Cloth Sci Technol 27(4):495–505. https://doi.org/10.1108/IJCST-05-2014-0065

Liu H, Chen D, Wei Q, Pan R (2013) An investigation into the bust girth range of pressure comfort garment based on elastic sports vest. J Text Inst 104(2):223–230. https://doi.org/10.1080/00405000.2012.714940

Morooka H, Fukuda R, Nakahashi M, Morooka H, Sasaki H (2005) Clothing pressure and wear feeling at under-bust part on a push-up type brassiere. Sen-I Gakkaishi 61(2):53–58

Thomassey S, Bruniaux P (2013) A template of ease allowance for garments based on a 3D reverse methodology. Int J Ind Ergon 43(5):406–416. https://doi.org/10.1016/j.ergon.2013.08.002

Tsujisaka T, Azuma Y, Matsumoto Y-I, Morooka H (2004) Comfort pressure of the top part of men's socks. Text Res J 74(7):598–602. https://doi.org/10.1177/004051750407400707

Wong ASW, Li Y, Zhang X (2004) Influence of fabric mechanical property on clothing dynamic pressure distribution and pressure comfort on tight-fit sportswear. Sen-I Gakkaishi 60(10):293–299. https://doi.org/10.2115/fiber.60.293

You F, Wang JM, Luo XN, Li Y, Zhang X (2002a) Garment's pressure sensation (1): subjective assessment and predictability for the sensation. Int J Cloth Sci Technol 14(5):307–316. https://doi.org/10.1108/09556220210446121

You F, Wang JM, Luo XN, Li Y, Zhang X (2002b) Garments pressure sensation (2): the psychophysical mechanism for the sensation. Int J Cloth Sci Technol 14(5):317–327. https://doi.org/10.1108/09556220210446248

Zeng X, Koehl L (2003) Representation of the subjective evaluation of the fabric hand using fuzzy techniques. Int J Intell Syst 18(3):355–366. https://doi.org/10.1002/int.10092

Zeng X, Ding Y, Koehl L (2004) A 2-tuple fuzzy linguistic model for sensory fabric hand evaluation. In: Ruan D, Zeng X (eds) Intelligent sensory evaluation. Springer, Berlin, Heidelberg, pp 217–234. https://doi.org/10.1007/978-3-662-07950-8_12

Zeng X, Ruan D, Koehl L (2008) Intelligent sensory evaluation: concepts, implementations, and applications. Math Comput Simulat 77(5–6):443–452. https://doi.org/10.1016/j.matcom.2007.11.013

Zhang X, Yeung K, Li Y (2002) Numerical simulation of 3D dynamic garment pressure. Text Res J 72(3):245–252. https://doi.org/10.1177/004051750207200311

Zhang M, Dong H, Fan X, Dan R (2015) Finite element simulation on clothing pressure and body deformation of the top part of men's socks using curve fitting equations. Int J Cloth Sci Technol 27(2):207–220. https://doi.org/10.1108/IJCST-12-2013-0139

Garment Fit Evaluation Using Machine Learning Technology

Kaixuan Liu, Xianyi Zeng, Pascal Bruniaux, Xuyuan Tao, Edwin Kamalha and Jianping Wang

Abstract Presently, garment fit evaluation mainly focuses on real try-on and rarely deals with virtual try-on. With the rapid development of e-commerce, there is a profound growth of garment purchases through the Internet. In this context, fit evaluation of virtual garment try-on is vital in the clothing industry. In this chapter, we propose a Naive Bayes-based model to evaluate garment fit. The inputs of the proposed model are digital clothing pressures of different body parts, generated from a 3D garment CAD software, while the output is the predicted result of garment fit (fit or unfit). To construct and train the proposed model, data on digital clothing pressures and garment real fit was collected for input and output learning data, respectively. By learning from these data, our proposed model can predict garment fit rapidly and automatically without any real try-on; therefore, it can be applied to remote garment fit evaluation in the context of e-shopping. Finally, the effectiveness of our proposed method was validated using a set of test samples. Test results showed that digital clothing pressure is a better index than ease allowance to evaluate garment fit, and machine learning-based garment fit evaluation methods have higher prediction accuracies.

Keywords Digital clothing pressure · Support vector machines · Naive Bayes
Active learning · Ease allowance · Real try-on

K. Liu (✉)
Apparel and Art Design College, Xi'an Polytechnic University, Xi'an 710048, China
e-mail: liukaixuan819@163.com

K. Liu · X. Zeng · P. Bruniaux · X. Tao · E. Kamalha
University of Lille 1, Nord de France, 59000 Lille, France

K. Liu · X. Zeng · P. Bruniaux · X. Tao · E. Kamalha
GEMTEX Laboratory, ENSAIT, 59100 Roubaix, France

J. Wang
College of Fashion and Design, Donghua University, Shanghai 200051, China

© Springer Nature Singapore Pte Ltd. 2018
S. Thomassey and X. Zeng (eds.), *Artificial Intelligence for Fashion
Industry in the Big Data Era*, Springer Series in Fashion Business,
https://doi.org/10.1007/978-981-13-0080-6_14

1 Introduction

Today garment e-shopping has become more prominent worldwide (Young Kim and Kim 2004). However, an important technical barrier that garments displayed online cannot be physically evaluated for fitting effects on a specific consumer (Song and Ashdown 2015). Thus, virtual try-on technology was developed to evaluate garment fit (Song and Ashdown 2015; Kim and Forsythe 2008), finding wide application in the clothing industry in the last ten years. A number of virtual try-on programs, such as *Clo 3D*, *Lectra 3D Prototype*, *OptiTex,* and *V-Stitcher 3D*, are available on the market for garment fit evaluation (Sayem et al. 2010). These 3D virtual try-on software systems follow similar principles, i.e., showing virtual garment static and dynamic performance from identified human morphological and fabric properties and their interactions. Defining this performance involves the use of complex mechanical and geometric modeling and simulation techniques, such as finite elements (Zhang et al. 2002). The software normally includes three main modules (Sayem et al. 2010): (1) a 3D parametric mannequin module, (2) a fabric properties module, and 3) a virtual pattern sewing module. To model the human body rapidly, the 3D parametric mannequin module is used to construct a personalized 3D human model from measurement of a 3D body scanner or a measuring tape, related to a specific customer. Several key body dimensions, such as height, waist circumference, and hip circumference (Liu et al. 2016b), control the parametric mannequin's dimensions. By adjusting key body dimensions, the 3D parametric mannequin module can create various body shapes and dimensions rapidly and automatically, meeting customers' body shapes and dimensions. Then, the fabric properties module, usually based on a mechanical model, will permit the simulation of different perceived properties (draping, texture, elasticity, bending, etc.) of a virtual fabric through adjustable fabric technical parameters. Finally, the virtual pattern sewing module assembles the predefined garment patterns on the specific 3D human body and sews the patterns together, taking into account the performance of the simulated fabric. The combination of these three modules constitutes a virtual try-on system, permitting the simulation of the real garment making process. In a virtual 3D try-on process, consumers and designers can visualize the static and dynamic performance of the selected fabrics and garment fit effects in terms of comfort, expressed by simulated pressures between the human body and fabrics, and fashion styles.

Through virtual try-on, consumers can easily decide whether they like a garment style or not. However, these virtual try-on applications are strongly dependent on mathematical models used in the software (Zhang et al. 2002) and cannot give full accurate garment fit evaluation. Moreover, for real try-on, the wearer can feel whether a garment fits, unlike, with virtual try-on (Shin and Baytar 2013). The issue of garment fit evaluation is a research hotspot and a great challenge (Fan et al. 2004). In practice, no matter how beautiful a garment is, and how excellent the fabric's properties are, a customer will not select it if it is unfit (Fan et al. 2004). Garment fit is a major factor affecting customers' purchasing decisions (Song and Ashdown 2012; Kamalha et al. 2013). For garment e-shopping, consumers cannot physically try-on garments; therefore, estimating the garment fit without real try-on is still an issue for researchers.

Lately, there are mainly two methods to evaluate garment fit through virtual try-on. One approach is that the visual evaluation is carried out on a 3D garment by expert fashion designers (Tao and Bruniaux 2013; Chen et al. 2015; Thomassey and Bruniaux 2013). Obviously, this subjective visual evaluation is neither accurate nor convincing. The other approach is to measure the ease allowances, which is the dimensional difference between human body and garment in the girth direction (Lee et al. 2007). Then, expert fashion designers analyze these measured ease allowances based on their own empirical knowledge to evaluate garment fit. However, the ease allowance can reflect the fitting feeling when neither it is less or equal to zero (tight garment style) nor does it take into account fabric properties. With the same value of ease allowance, the fitting effects could be different if the fabrics are different. Evidently, ease allowance only is not enough for characterizing the fitting effects of a garment try-on. Moreover, these two approaches to garment fit need empirical knowledge. Often, the predicted results are entirely dependent on the subjectivity of designers, such as experience and personal preference, which erodes their accuracy. Therefore, people without fashion design knowledge cannot easily use the traditional fit evaluation methods of virtual garments. For shopping online, there are thousands of garments purchased in a short time. If every garment fit is evaluated using traditional methods, the work is so enormous. Therefore, it is necessary to find a method that can evaluate garment fit automatically, rapidly, and accurately. In this context, we proposed a machine learning-based model to evaluate garment fit. The input item of the proposed model is an indicator reflecting the garment fit condition, whose output is fit or unfit. Compared to traditional garment fit evaluation methods, the greatest advantage of our proposed method is that it can predict garment fit rapidly and automatically, without any real try-on and designers' involvement.

Modeling by learning from experimental data has been widely used in the clothing industry (Guo et al. 2011; Liu et al. 2017), including evaluation of garment and fabric products (Chen et al. 2015; Zeng and Koehl 2003; Zeng and Liu 2005; Zeng et al. 2008), evaluation of wear comfort (Wong et al. 2004), garment CAD systems (Kim and Cho 2000; Vuruskan et al. 2015; Wang et al. 2015; Hu et al. 2008), clothing manufacturing (Wong et al. 2005; Guo et al. 2008; Lin 2009; Hui et al. 2002; Guo et al. 2009; Çiflikli and Kahya-Özyirmidokuz 2010), clothing retailing (Sun et al. 2008; Wong and Guo 2010; Wong et al. 2009; Choi et al. 2015; Xia et al. 2012; Xia and Wong 2014), and apparel supply chain management (Lo et al. 2008; Pan et al. 2009). However, few studies have focused on fit evaluation of virtual try-on using machine learning. In order to increase the accuracy of fit evaluation using 3D virtual body shapes and virtual garments, we introduce a data machine learning-based model. This method requires inputting an indicator that can reflect garment fit and returning an output as the predicted result of garment fit. In the preceding sections, we have discussed and indicated that the ease allowance between a garment and the human body is not a good indicator of reflecting garment fit. We, therefore, opted to find a more suitable indicator. The influence of fabric properties can be measured using the digital clothing pressures, distributed over the human body of the wearer and provided by a 3D garment CAD software like *CLO 3D* (Liu et al. 2016a, c, 2017). It is for this reason that, in our study, we selected digital clothing pressures as a key indicator for remote garment fit prediction. In practice, the digital pressure-based method will be more efficient for fit evaluation than ease allowance-based methods, which are more adapted to loose garments.

Bayes classifiers have been applied successfully in a wide variety of domains (Langley et al. 1992). Their representations are quite intuitive and easy to understand (Domingos and Pazzani 1997). The advantages of Naive Bayes are (1) A Naive Bayes model has a solid mathematical foundation, as well as the stability of the classification efficiency. (2) The number of estimated parameters for modeling with Naive Bayes is relatively fewer, the model is less sensitive to missing data, and the algorithm is relatively simpler. (3) A Naive Bayes model has very high classification accuracy in many practical cases. Due to these advantages, Naive Bayes is applied to model the relationship between digital clothing pressures and garment fit level. The inputs of the proposed model are digital clothing pressures on different body parts, while the output is the fit evaluation result (fit or unfit). By learning from a number of experimental data measured on a number of samples, we set up the model, capable of quickly estimating the fit for a new garment without any real try-on. Our proposed model can be used to help consumers in realizing online efficient garment shopping.

The following sections are organized as follows. Section 2 introduces the general scheme and data formalization. Sections 3 presents the collection of input and output learning data, respectively. Section 4 expounds the construction of garment fit prediction model. In Sect. 5, we evaluate the accuracy of the proposed model and give a practical application of the proposed model. In Sect. 6, we discuss the application prospect, limitation, etc. Finally, we present some conclusions and possible further works in Sect. 7.

2   General Principle and Formalization

2.1   General Principle

The general scheme of the mentioned garment fit evaluation model is described in Fig. 1.

First, Experiment I is aimed at collecting the output learning data. Nine subjects try on 72 pairs of pants, respectively. They divide all the pants into fit pants and unfit pants.

Second, Experiment II is aimed at collecting the input learning data. We measured digital clothing pressures of the 72 pairs of pants, respectively, by virtual try-on technology.

Next, garment fit evaluation model based on Naive Bayes and support vector machines (SVMs) is trained by the input and output learning data.

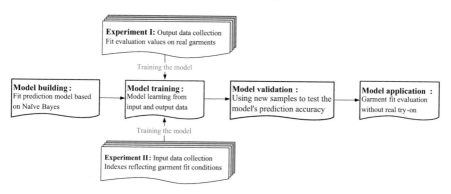

Fig. 1 General scheme of MLBGFET

Finally, the proposed model predicts garment fit without any real try-on after learning from the collected data.

2.2 Formalization of the Concepts and Data

As shown in Fig. 2, we built a machine learning-based garment fit evaluation model, the input of which are digital clothing pressures and the output of which is garment fit level. The data and concepts involved in this study are formalized as follows:

Let FL be the fit level of a garment, i.e., 1-fit, 0-unfit.

Let $G = \{g_1, g_2, \ldots, g_m\}$ be a set of m real garments used in our study.

Let $P_i = (p_i^1, \ldots, p_i^j, \ldots, p_i^k)$ be a vector of digital clothing pressures obtained during the virtual try-on of the garment g_i where p_i^j is the pressure on the key position j of the garment g_i (we suppose that there exist k key positions on the whole garment surface). In a general case, the vector of digital pressures $P_{new} = (p_{new}^1, \ldots, p_{new}^j, \ldots, p_{new}^k)$ of a new garment g is taken as input variables of the model.

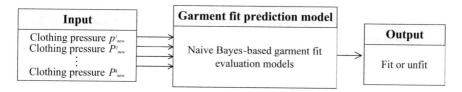

Fig. 2 Modeling the relation between digital clothing pressures and garment fit

3   Learning Data Acquisition

3.1   Preparation Work for Experiments

We design Experiments I and II to collect data. Experiment I aims to acquire output learning data on garment fit by using real try-on; Experiment II aims to acquire input learning data on digital clothing pressures by using virtual try-on. Anthropometric equipment, software, subjects, garments, fabrics, etc., involved in Experiment I and II are expounded below, respectively.

Anthropometric equipment: The Vitus Smart 3D body scanner is applied to collect human body dimensions for virtual try-on. This device captures body measurements with a ±1 mm level of accuracy, in accordance with the international standard DIN EN ISO 20685.

Software: The software CLO 3D is applied to measure digital clothing pressures. This software permits to create virtual, close-to-life garment visualization with cutting-edge simulation technologies. Virtual fabrics available in CLO 3D are based on actual fabrics commonly used in the industry, and they currently have a 95% accuracy rate (Enterprice 2016).

Subjects: Nine female subjects with representative body shapes are selected for performing real try-on and body dimension measurement. According to China National Standard (GBT 1335.2-2008), their body dimensions (155/60A, 155/62A, 160/64A, 160/66A, 160/68A, 165/70A, 165/72A, 170/74A, and 170/76A) can account for the total female population of China (Committee CNSM 2008). (Note: In China, female body shapes are classified into four categories (Y, A, B, C) according to the difference of bust–waist. The body shape belongs to the type Y if this value is located in the range of 19–24 cm, the type A for the range of 14–18 cm, the type B for the range of 9–13 cm, and the type C for the range of 4–8 cm. 155/60A means that the body type is A, the stature 155 cm and the waist 71 cm).

Garments: 72 pairs of straight pants, which cover most of pants' sizes, are involved in the real try-on experiments for data collection. We select the pant type to test our proposed method because that they are the most challenging clothing item for a good fit (Song and Ashdown 2015). If the proposed model predicts pants' fit accurately, this method could be also available for other styles.

Fabric: Fabric physical properties influence digital clothing pressures significantly. Therefore, they should be considered in the virtual try-on experiment. However, these fabric properties, as well as garment styles, have already been taken into account in the digital clothing pressures measured in the 3D garment CAD environment. Therefore, we do not need to specially study the effects of fabric properties on garment fit. In Experiment II, we just selected a frequently used jeans fabric with the mechanical properties shown in Table 1 for making different garments of virtual try-on.

Garment fit level: In this research, we classify all garment fit values into two levels fit or unfit. These fit levels are used in both real and virtual garment try-on.

Table 1 Values of the fabric mechanical properties

Buckling stiffness-weft	Buckling stiffness-warp	Buckling ratio-weft	Internal damping	Friction coefficient	Buckling ratio-warp
30	30	50	1	3	50
Bending-weft	Bending-warp	Shear	Stretch-weft	Stretch-warp	Density
35	35	23	32	32	35

Try-on condition: Before each real try-on evaluation, each subject wears a piece of underwear that is thin and neither tight nor loose. The try-on experiment is carried out indoor under a temperature of 18–20 °C.

3.2 Experiment I: Acquisition of the Data on Garment Fit

Experiment I is designed to evaluate garment fit levels using the real try-on. The experiment procedure is shown in Fig. 3. Nine selected female subjects with different body shapes participate in the garment fit evaluation procedure. The details are given below.

Step 1: Each of the nine subjects selects eight pants from the 72 pairs of real pants according to her personal preference, like what she usually does in a garment shop. One pair of pants is selected by only one subject.

Step 2: Each subject realizes her try-on with the selected pants by performing a number of gestures: sitting down, standing, squatting, running, and walking (see Fig. 3). After that, she gives an overall fit level of the evaluated pants using one of the two scores (fit or unfit).

Finally, the nine subjects evaluate the fit levels of all the 72 pairs of pants. According to these evaluation results, the 72 pairs of pants are classified into the set of fit pants (23 pairs) and the set of unfit pants (49 pairs) (see Table 2). The data will be taken as input learning data to build the proposed models.

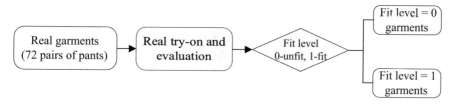

Fig. 3 Garment fit data collection by real try-on (output learning data)

Table 2 Garment fitness data collected by real try-on (output learning data)

Fit levels	Fit	Fit	Fit	Fit	Unfit	Unfit	Unfit	Unfit
Sample no.	1	2	…	23	24	25	…	72

3.3 Experiment II: Acquisition of the Data on Digital Clothing Pressures

We design Experiment II to measure the digital clothing pressures at the key positions of the garment surface using the *CLO 3D* software (Fig. 4). The concrete scheme of Experiment II is described as follows.

Step 1: We built nine 3D human models whose body dimensions are equal to those of the nine subjects.

Step 2: We determine key positions $F1, F2, …, F15$ and $B1, B2, …, B5$ of each pair of pants, which are uniformly distributed on the front piece pattern and on the back piece pattern, respectively (Fig. 4a). As the parts below knee have little effect on clothing fit, we do not define any key positions on them.

Step 3: We make virtual try-on with the patterns of the 72 pairs of pants on the 3D human models corresponding to the body dimensions of nine subjects previously selected (Fig. 4b).

Step 4: We measure the digital clothing pressures of each pair of pants on predefined 20 key positions of each garment during its virtual try-on (Fig. 4c).

The digital clothing pressure data of the 72 pairs of pants were collected. The corresponding data (input data) will be combined with the data of garment fit evaluation, collected in Sect. 3.2 (output data), for building the fit prediction models.

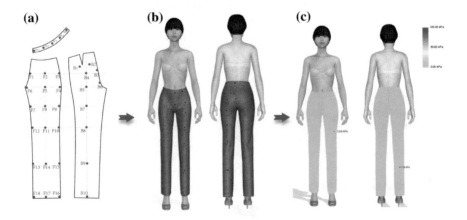

Fig. 4 Digital clothing pressure measurement by virtual try-on (input learning data)

4 Modeling the Relation Between Clothing Pressures and Garment Fit Level

As an effective tool for modeling with data learning, the Naive Bayes classifier is used in our approach for constructing the garment fit evaluation model. We present the specific modeling procedure in Fig. 5. It is composed of the following five steps:

Step 1: *Determining the characteristics of attributes*
In the procedure of modeling, the digital clothing pressures that measure on the k predefined key positions on the garment are taken as the characteristics attributes of the model. According to the general principle of Naive Bayes, we suppose that these k characteristics attributes are independent each other and all respect normal distributions.

Step 2: *Acquiring training samples*
Two experiments I and II are carried out to collect training data by real and virtual try-on. The input and output training data are digital clothing pressures and garment fit levels, respectively.

Step 3: *Computing the prior probabilities of each category*
We have two levels of garment fit (categories). Thus, the prior probability of each category $i (i \in \{1, 0\})$ can be:

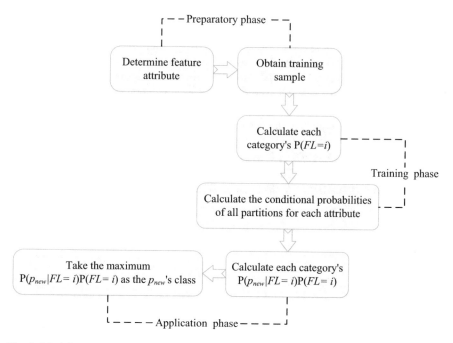

Fig. 5 Modeling with Naive Bayes

$$P(FL_i) = \frac{\text{the number of fit evalutions corresponding to the } i\text{-the level}}{\text{the total number of fit evaluations}}$$

Step 4: *Computing the conditional probability of the new sample P_{new}*

$$P(P_{new}|FL = i) = \prod_{j=1}^{k} P(P_{new}^{j}/FL = i) \; (i \in \{1, 0\})$$

$$P(FL = i|P_{new}) = \frac{P(FL = i)P(P_{new}|FL = i)}{\sum_{i=1}^{5} P(FL = i)P(P_{new}|FL = i)}(i\{1, 0\})$$

Step 5: *Predicting with Naive Bayes classifier.*

The classification rule of the Naive Bayes classifier is given below.

If $P(FL = l|P_{new}) = \max_{i=1 \text{ or } 0}\{P(FL = i|P_{new})\}$, $(l \in \{1, 0\})$, then the new sample P_{new} corresponds to the fit level l.

5 Model Validation

In this section, we compare the performances of the proposed model. As the number of learning data is very limited in this research, which may cause "over fitting" problem, leading to a very unstable performance of the model output, we apply K-fold cross-validation approach to calculate the prediction accuracies of the proposed model. To compare with other machine learning algorithms, we also calculate the prediction accuracies of the SVMs-based and ease allowance-based garment fit evaluation models. According to the definition of ease allowance, the waist's ease allowance equals to garment's waist girth minus human body's waist girth, and the hip's ease allowance equals to garment's hip girth minus human body's hip girth.

The test result indicates that the prediction accuracy of Naive Bayes model (93.1%) is better than that of SVMs model (84.7%) with selected digital clothing pressure as a fit evaluation index. The prediction accuracy of Naive Bayes model (76.4%) is slightly worse than that of SVMs model (77.8%) with selected ease allowance as a fit evaluation index.

6 Discussion

6.1 Influence of the Difference Between Real and Digital Pressures on the Prediction Results

In the introduction, we have pointed out that the ease allowance between a garment and the human body is not a good indicator of reflecting garment fit. Therefore, we opted to find a more suitable indicator. The influence of fabric properties can be measured using the digital clothing pressures distributed on the garment surface covering the human body of the wearer. These digital clothing pressures are easily measured in a garment CAD software environment like CLO 3D. Our previous research shows that the digital clothing pressures can reflect garment wear comfort accurately (Liu et al. 2016a). It is for this reason that we select the digital clothing pressures as a key indicator for performing remote garment fit prediction without real try-on. The test result indicates that the digital pressure-based methods are more efficient in fit evaluation than the ease allowance-based methods, which can be adapted to loose garments only instead of tight ones.

The proposed models enable to set up accurate and quantitative relations between digital clothing pressure data measured during a virtual try-on and garment fit data evaluated during a real try-on. For a new garment with an unknown fit level, we can measure its digital clothing pressures and then apply a previously proposed model for predicting its fit according to the measured digital clothing pressures. These models are significant and can accurately reflect comfort feeling of garments with different fabric mechanical properties because the digital and real clothing pressures not only have the same variation trends (i.e., the digital clothing pressure at a position is high when a subject feels tight at the same position, and vice versa.) (Zhang et al. 2002; Liu et al. 2016a, c; Seo et al. 2007; Yanmei et al. 2014; Zhang et al. 2015), but also are rather close each other in a certain range. As the learning and prediction of the proposed model are both based on digital clothing pressures and do not deal with any real clothing pressures, we do not need to precisely identify real clothing pressures. Even if there are some differences between digital and real clothing pressures, the prediction accuracies of the proposed models are not affected.

6.2 Application Prospect

With increasing online sales, the fit of garments has serious implications for a fashion retailer because ill-fitting garments are directly related to product return rates (Des-Marteau 2000; Kim and Damhorst 2010). The evaluation of garment fit without the physical participation of customers and designers is very useful for online clothing shoppers. In this context, we introduce an application based on the proposed method to predict garment fit in an e-shopping environment (Fig. 6).

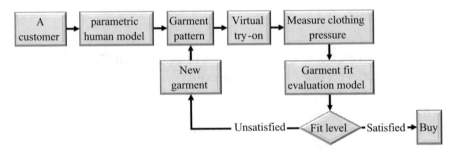

Fig. 6 Remote garment fit prediction for online shopping

As shown in Fig. 6, for a specific customer, a parametric human model is used to be adapted or adjusted to the real dimensions of the concerned human body; next, we search for the garment patterns from the database of the company according to the previous body dimensions; next, a number of red points will be marked on the selected patterns in order to measure the clothing pressures at these key positions; then, garment patterns are assembled on the adjusted digital human model; next, the assembled patterns are seamed together to form a 3D virtual garment; digital clothing pressures are measured on the predefined key positions finally.

Having performed the previous operations, the collected digital clothing pressures are introduced to the garment fit evaluation model (see the red wireframe in Fig. 6) for predicting the garment fit automatically. If the predicted result meets the customer's requirement, we recommend the concerned customer to buy the garment. Otherwise, she/he will be invited to try another one with a different size or style. This procedure repeats until the satisfaction of the result.

6.3 Limitation and Future Research

The limitations and future research are summarized as follows:

(1) In order to get reliable data to train the proposed method, garments with different sizes and styles need to be made first. As garment patterns are the business secret for fashion companies, we needed to make garment patterns and real garments by ourselves. Due to this reason, we only collected a small dataset with 72 samples to train the proposed model. Thus, further research can be combined with a specific garment company to train the proposed model using their existing clothing. As garment patterns are the business secret, our proposed approach might be only suitable for companies do both production and sale of garments by themselves.

(2) The two fit levels (fit and unfit) are too simplified. For example, during the real try-on, we only evaluate the overall fit level for all the gestures and all the positions. In fact, more accurate results can be obtained if we propose to evaluate

a series of local fit levels (hip fit level, waist fit level, etc.) each corresponding to one body position of the wearer and then properly aggregate them for generating an overall fit level. In this situation, all local discomfort feeling can also be considered in the fit prediction models in the further research.

(3) Digital clothing pressure is selected as the index of garment fit evaluation. However, it is possible that some parts of a loose garment are not in contact with the human body and the corresponding clothing pressures could be near zero. In further research, ease allowance and clothing pressure can be combined together for evaluating garment fit in a complementary way.

(4) The used digital clothing pressures are static values measured at different key positions related to a given gesture. The dynamic aspect, i.e., the clothing pressures varying with time during a movement is not considered. We need to apply time series analysis to study these clothing pressures and form new input variables of the fit evaluation model in further research.

7 Conclusion

In this research, we proposed a machine learning-based model to predict garment fit. The results indicate that: (1) digital clothing pressure is a better garment evaluation index than ease allowance; (2) Naive Bayes is a good classifier and not inferior to other classifiers in the field of garment fit evaluation, and even better than SVMs in some cases. Compared with the traditional garment fit evaluation methods, the proposed approach has a number of advantages: (1) continuous improvement of the model's performance with new learning data, (2) independence of any real try-on, (3) removal of human involvement. Due to the dataset is very small in this research, the approach might not be easy to use in a real-life scenario. More data should be collected in the future practical applications.

Acknowledgements This research was financially supported by China National Endowment for the Arts.

References

Chen X, Tao X, Zeng X, Koehl L, Boulenguez-Phippen J (2015) Control and optimization of human perception on virtual garment products by learning from experimental data. Knowl-Based Syst 87:92–101. https://doi.org/10.1016/j.knosys.2015.05.031
Choi S, Yang Y, Yang B, Cheung H (2015) Item-level RFID for enhancement of customer shopping experience in apparel retail. Comput Ind 71:10–23. https://doi.org/10.1016/j.compind.2015.03.003
Çiflikli C, Kahya-Özyirmidokuz E (2010) Implementing a data mining solution for enhancing carpet manufacturing productivity. Knowl-Based Syst 23(8):783–788. https://doi.org/10.1016/j.knosys.2010.05.001

Committee CNSM (2008) GBT 1335.2-2008. Standard sizing systems for garments. Standards Press of China, Beijing

DesMarteau K (2000) CAD: let the fit revolution begin. Bobbin 42(2):42–56

Domingos P, Pazzani M (1997) On the optimality of the simple Bayesian classifier under zero-one loss. Mach Learn 29(2):103–130. https://doi.org/10.1023/a:1007413511361

Enterprice CD (2016) CLO 3D. https://www.clo3d.com/

Fan J, Yu W, Hunter L (2004) Clothing appearance and fit: science and technology. Woodhead publishing Limited, Cambridge, UK

Guo ZX, Wong WK, Leung SYS, Fan JT, Chan SF (2008) Genetic optimization of order scheduling with multiple uncertainties. Expert Syst Appl 35(4):1788–1801. https://doi.org/10.1016/j.eswa. 2007.08.058

Guo ZX, Wong WK, Leung SYS, Fan JT (2009) Intelligent production control decision support system for flexible assembly lines. Expert Syst Appl 36 (3, Part 1):4268–4277. https://doi.org/ 10.1016/j.eswa.2008.03.023

Guo Z, Wong W, Leung S, Li M (2011) Applications of artificial intelligence in the apparel industry: a review. Text Res J 81(18):1871–1892. https://doi.org/10.1177/0040517511411968

Hu Z-H, Ding Y-S, Zhang W-B, Yan Q (2008) An interactive co-evolutionary CAD system for garment pattern design. Comput-Aided Des 40(12):1094–1104. https://doi.org/10.1016/j.cad.2008. 10.010

Hui PCL, Chan KCC, Yeung KW, Ng FSF (2002) Fuzzy operator allocation for balance control of assembly lines in apparel manufacturing. IEEE T Eng Manag 49(2):173–180. https://doi.org/10. 1109/TEM.2002.1010885

Kamalha E, Zeng Y, Mwasiagi JI, Kyatuheire S (2013) The comfort dimension; a review of perception in clothing. J Sens Stud 28(6):423–444. https://doi.org/10.1111/joss.12070

Kim H-S, Cho S-B (2000) Application of interactive genetic algorithm to fashion design. Eng Appl Artif Intel 13(6):635–644. https://doi.org/10.1016/S0952-1976(00)00045-2

Kim H, Damhorst ML (2010) The relationship of body-related self-discrepancy to body dissatisfaction, apparel involvement, concerns with fit and size of garments, and purchase intentions in online apparel shopping. Cloth Text Res J 28(4):239–254. https://doi.org/10.1177/0887302x10379266

Kim J, Forsythe S (2008) Adoption of virtual try-on technology for online apparel shopping. J Interact Mark 22(2):45–59. https://doi.org/10.1002/dir.20113

Langley P, Iba W, Thompson K (1992) An analysis of Bayesian classifiers. In: AAAI'92 Proceedings of the tenth national conference on artificial intelligence, Menlo Park, CA. AAAI Press, pp 223–228

Lee J, Nam Y, Cui MH, Choi KM, Choi YL (2007) Fit evaluation of 3D virtual garment. In: Aykin N (ed) Usability and internationalization. HCI and Culture. Springer, Berlin, pp 550–558. https://doi.org/10.1007/978-3-540-73287-7_64

Lin M-T (2009) The single-row machine layout problem in apparel manufacturing by hierarchical order-based genetic algorithm. Int J Cloth Sci Technol 21(1):31–43. https://doi.org/10.1108/ 09556220810898872

Liu K, Kamalha E, Wang J, Agrawal T-K (2016a) Optimization design of cycling clothes' patterns based on digital clothing pressures. Fiber Polym 17(9):1522–1529. https://doi.org/10.1007/ s12221-016-6402-2

Liu K, Wang J, Zeng X, Tao X, Bruniaux P, Edwin K (2016b) Fuzzy classification of young women's lower body based on anthropometric measurement. Int J Ind Ergon 55(5):60–68. https://doi.org/ 10.1016/j.ergon.2016.07.008

Liu K, Wang J, Zhu C, Hong Y (2016c) Development of upper cycling clothes using 3D-to-2D flattening technology and evaluation of dynamic wear comfort from the aspect of clothing pressure. Int J Cloth Sci Technol 28(6):736–749. https://doi.org/10.1108/IJCST-02-2016-0016

Liu K, Wang J, Kamalha E, Li V, Zeng X (2017a) Construction of a body dimensions' prediction model for garment pattern making based on anthropometric data learning. J Text Inst 108(12):2107–2114. https://doi.org/10.1080/00405000.2017.1315794

Liu K, Wang J, Hong Y (2017b) Wearing comfort analysis from aspect of numerical garment pressure using 3D virtual-reality and data mining technology. Int J Cloth Sci Technol 29(2):166–179. https://doi.org/10.1108/IJCST-03-2016-0017

Lo W-S, Hong T-P, Jeng R (2008) A framework of E-SCM multi-agent systems in the fashion industry. Int J Prod Econ 114(2):594–614. https://doi.org/10.1016/j.ijpe.2007.09.010

Pan A, Leung S, Moon K, Yeung K (2009) Optimal reorder decision-making in the agent-based apparel supply chain. Expert Syst Appl 36(4):8571–8581. https://doi.org/10.1016/j.eswa.2008.10.081

Sayem ASM, Kennon R, Clarke N (2010) 3D CAD systems for the clothing industry. Int J Fash Des Technol Educ 3(2):45–53. https://doi.org/10.1080/17543261003689888

Seo H, Kim S-J, Cordier F, Hong K (2007) Validating a cloth simulator for measuring tight-fit clothing pressure. In: Proceedings of the 2007 ACM symposium on solid and physical modeling, Beijing, China, 2007. ACM, 1236308, pp 431–437. https://doi.org/10.1145/1236246.1236308

Shin E, Baytar F (2013) Apparel fit and size concerns and intentions to use virtual try-on: impacts of body satisfaction and images of models' bodies. Cloth Text Res J 32(1):20–33. https://doi.org/10.1177/0887302x13515072

Song HK, Ashdown SP (2012) Development of automated custom-made pants driven by body shape. Cloth Text Res J 30(4):315–329. https://doi.org/10.1177/0887302x12462058

Song HK, Ashdown SP (2015) Investigation of the validity of 3-D virtual fitting for pants. Cloth Text Res J 33(4):314–330. https://doi.org/10.1177/0887302X15592472

Sun Z-L, Choi T-M, Au K-F, Yu Y (2008) Sales forecasting using extreme learning machine with applications in fashion retailing. Decis Support Syst 46(1):411–419. https://doi.org/10.1016/j.dss.2008.07.009

Tao X, Bruniaux P (2013) Toward advanced three-dimensional modeling of garment prototype from draping technique. Int J Cloth Sci Technol 25(4):266–283. https://doi.org/10.1108/09556221311326301

Thomassey S, Bruniaux P (2013) A template of ease allowance for garments based on a 3D reverse methodology. Int J Ind Ergon 43(5):406–416. https://doi.org/10.1016/j.ergon.2013.08.002

Vuruskan A, Ince T, Bulgun E, Guzelis C (2015) Intelligent fashion styling using genetic search and neural classification. Int J Cloth Sci Technol 27(2):283–301. https://doi.org/10.1108/IJCST-02-2014-0022

Wang LC, Zeng XY, Koehl L, Chen Y (2015) Intelligent fashion recommender system: fuzzy logic in personalized garment design. IEEE Trans Human Mach Syst 45(1):95–109. https://doi.org/10.1109/THMS.2014.2364398

Wong W, Guo Z (2010) A hybrid intelligent model for medium-term sales forecasting in fashion retail supply chains using extreme learning machine and harmony search algorithm. Int J Prod Econ 128(2):614–624. https://doi.org/10.1016/j.ijpe.2010.07.008

Wong A, Li Y, Yeung P (2004) Predicting clothing sensory comfort with artificial intelligence hybrid models. Text Res J 74(1):13–19. https://doi.org/10.1177/004051750407400103

Wong W, Leung S, Au K (2005) Real-time GA-based rescheduling approach for the pre-sewing stage of an apparel manufacturing process. Int J Adv Manuf Tech 25(1–2):180–188. https://doi.org/10.1007/s00170-003-1819-3

Wong WK, Zeng X, Au W, Mok P, Leung S (2009) A fashion mix-and-match expert system for fashion retailers using fuzzy screening approach. Expert Syst Appl 36(2):1750–1764. https://doi.org/10.1016/j.eswa.2007.12.047

Xia M, Wong WK (2014) A seasonal discrete grey forecasting model for fashion retailing. Knowl-Based Syst 57:119–126. https://doi.org/10.1016/j.knosys.2013.12.014

Xia M, Zhang Y, Weng L, Ye X (2012) Fashion retailing forecasting based on extreme learning machine with adaptive metrics of inputs. Knowl-Based Syst 36:253–259. https://doi.org/10.1016/j.knosys.2012.07.002

Yanmei L, Weiwei Z, Fan J, Qingyun H (2014) Study on clothing pressure distribution of calf based on finite element method. J Text Inst 105(9):955–961. https://doi.org/10.1080/00405000.2013.865883

Young Kim E, Kim YK (2004) Predicting online purchase intentions for clothing products. Eur J Marketing 38(7):883–897. https://doi.org/10.1108/03090560410539302

Zeng X, Koehl L (2003) Representation of the subjective evaluation of the fabric hand using fuzzy techniques. Int J Intell Syst 18(3):355–366. https://doi.org/10.1002/int.10092

Zeng X, Liu Z (2005) A learning automata based algorithm for optimization of continuous complex functions. Inf Sci 174(3–4):165–175. https://doi.org/10.1016/j.ins.2004.09.004

Zeng X, Ruan D, Koehl L (2008) Intelligent sensory evaluation: concepts, implementations, and applications. Math Comput Simulat 77(5–6):443–452. https://doi.org/10.1016/j.matcom.2007.11.013

Zhang X, Yeung K, Li Y (2002) Numerical simulation of 3D dynamic garment pressure. Text Res J 72(3):245–252. https://doi.org/10.1177/004051750207200311

Zhang M, Dong H, Fan X, Dan R (2015) Finite element simulation on clothing pressure and body deformation of the top part of men's socks using curve fitting equations. Int J Cloth Sci Technol 27(2):207–220. https://doi.org/10.1108/IJCST-12-2013-0139

Erratum to: Artificial Intelligence for Fashion Industry in the Big Data Era

Sébastien Thomassey and Xianyi Zeng

Erratum to:
S. Thomassey and X. Zeng (eds.), *Artificial Intelligence*
for Fashion Industry in the Big Data Era,
Springer Series in Fashion Business,
https://doi.org/10.1007/978-981-13-0080-6

In the original version of the book, the following corrections have to be incorporated:

In chapter "AI-Based Fashion Sales Forecasting Methods in Big Data Era", the affiliation "Institute of Textiles and Clothing, The Hong Kong Polytechnic University, Hunghom, Kowloon, Hong Kong" of the co-author "Shuyun Ren" has to be changed as "Guangdong University of Technology, China".

In chapter "Blockchain-Based Secured Traceability System for Textile and Clothing Supply Chain", the incorrect affiliation "Independent consultant, Chennai, India" of the co-author "Ajay Sharma" has to be corrected so that it should read as "Independent Consultant, Roubaix, France".

The updated online version of these chapters can be found at
https://doi.org/10.1007/978-981-13-0080-6_2
https://doi.org/10.1007/978-981-13-0080-6_10
https://doi.org/10.1007/978-981-13-0080-6

ated in the United States
Bookmasters